ECOLOGICAL
ETHICS
AND
POLITICS

PHILOSOPHY AND SOCIETY
General Editor: Marshall Cohen

Also in this series:

ECOLOGICAL ETHICS AND POLITICS

H. J. McCloskey

ROWMAN AND LITTLEFIELD
Totowa, New Jersey

First published in the United States 1983 by Rowman and Littlefield, 81 Adams Drive, Totowa, New Jersey 07512.

Library of Congress Cataloging in Publication Data

McCloskey, Henry John, 1925-
  Ecological ethics and politics.

  (Philosophy and society)
  Bibliography: p.
  1. Human ecology—Moral and ethical aspects.
2. Environmental protection—Moral and ethical
aspects. 3. Environmental policy. I. Title.
II. Series.
GF80.M37   304.2   82-3840
ISBN 0-8476-7111-9   AACR2

Printed in the United States of America

Contents

v

Acknowledgments

In preparing this book, I have developed material from two of my published papers, namely, "Ecological Ethics and Its Justification: A Critical Appraisal" (1980a) and "Ecological Values, the State and the Right to Liberty" (1980b).

Many people have contributed to the production of this work. John Passmore, through his book, *Man's Responsibility For Nature*, and in discussions at the Australian National University, did much to make me aware that ecological ethics and politics were important areas of philosophy. An invitation by Don Mannison of the University of Queensland to participate in a conference on environmental ethics at that university in 1977, which I accepted, and discussions at that conference, most especially with Richard Routley, did much to decide me to direct my research and writing toward that area. Professor Preston King of the School of Politics, University of New South Wales, invited me to present a paper to a staff seminar on issues discussed in Part III. I gained much from the discussion that followed the paper and thank those who contributed for the insights I gained from that seminar. I have learned much by way of discussions with colleagues at La Trobe University, most especially from Ms. Janna Thompson and Dr. Robert Young, and through being able to participate in a course in environmental philosophy planned and organized by Ms. Thompson.

My greatest debts are to my friends Dr. Jan Srzednicki of the Philosophy Department, University of Melbourne, and Mr. Patrick Singleton, Baillieu Library, University of Melbourne. Dr. Srzednicki read two drafts of this work and made copious, careful, critical, and constructive comments, which led to a rethinking of and a new approach to various of the issues. Mr. Singleton gave very generously of his knowledge and time in assisting me to come to know and to take account of relevant material of which otherwise I should have had no knowledge. Dr. Mary A. McCloskey made many helpful suggestions during the writing of the book and suggested various minor revisions to the final draft. My thanks also go to Suzanne G.

Hayster for typing the work and for her cooperative help during the writing of the book. Mr. John Myhill greatly assisted me in the final editing of the work immediately before sending it to press.

La Trobe University
1982

PART I

Ecology and Its Relevance to Moral and Political Concerns

1

Introduction

Ecology is explained as being a branch of biology dealing with living organisms' habits, modes of life, and relations to their surroundings. As such, ecology would appear to be a significant area of science, one that has obvious relevance to our practical concerns, both day to day and long term, although of a kind not so immediately evident in respect of the more theoretical, abstract sciences. Such an account gives little hint of the vast practical importance that has been attributed to the findings of ecology in recent years. Many attribute to ecology something like the importance formerly accorded the discovery of Newtonian physics, urging that our failure to take note of the discoveries of ecology is hurtling us toward "an ecological crisis," a crisis that threatens not simply the quality of life to be enjoyed by human beings but the very survival of the human race; that ecology reveals that *Homo sapiens* is, or is in danger of becoming, an endangered species; and more, that man is "ravaging" the earth in a reckless, irresponsible way, doing "irreversible damage" to it and thereby rendering it of less value in itself and for its members, including mankind. Thus it is contended that only in the light of the findings of ecology can we determine what is in our short- and long-term self-interest, what are our personal and social moral obligations, what are the responsibilities of the state and the political organizations of the world regarding man's use of the earth and its constituents. During the past thirty years a veritable industry has grown up around the findings of ecologists, with numerous writings directed at showing the importance of the truths of ecology for the pursuit of one's self-interest, the self-interest of society, the self-interest of the human race, the well-being of nature, as well as for determining our moral duties toward our fellow human beings, the future generations, and Nature generally, and concerning the kinds of social and political arrangements that will be necessary.

So great has been the outpouring of writings, books popular and learned, articles in learned journals and popular magazines, newspaper reports, TV discussions and documentaries, and the like that they themselves now have a significant ecological impact on the use of raw materials, resources such as trees, fossil fuels, mercury and other nonrenewable resources, the production of pollution, encroachments on wilderness, and the further endangering of species. The truths of ecology need to be both basic and of great importance in order for this great activity and clamor to be justified, and in order for the demand for a basic rethinking of what is dictated by personal, social, and species prudence and personal, social, and political morality to be as well grounded as it is so widely claimed to be.

The major contributions of ecology relate to its stress on the interconnectedness of things in nature, in and through ecosystems and between ecosystems, as in food chains and biotic pyramids, and on the delicacy of the balances, equilibriums, integrities of the various ecosystems, and of the world ecosystem, such that these equilibriums may be disturbed, ecosystems destroyed, intentionally or unintentionally, with effects that are harmful to man and the environment in which he lives. Ecology reveals that we are in danger of bringing about changes and states of affairs that we should wish to avoid, and that we can avoid, if we act quickly, decisively, and on the basis of ecological knowledge and practical ecological wisdom. While different theorists assess the nature, urgency, and gravity of the problems differently, there is a common concern about the harm, much of it needless, unintended, unforeseen, and irreversible, that man is doing.

Thus we find theorists making claims about the relevance and importance of ecology, of which the following are a sample: "History will show, I think, that the two most important dates in biology in the 20th century (so far) are 1953, when Watson and Crick published their paper on the structure of DNA; and 1962, when Rachel Carson published *Silent Spring*. The importance of the former is universally recognized; of the latter not" (Hardin, 1969, 151). Hardin has also observed: "The world is an unimaginably complex system of systems; if we are to live well, this is what we must try to understand" (1975). "Man will ultimately destroy himself if he thoughtlessly eliminates the organisms that constitute essential links in the complex and delicate web of life of which he is a part" (Dubos, 1970, 15). "Man-made changes in the biosphere threaten the integrity of the life-support system essential for the survival of human life . . . The environmental problem, on the other hand, is frequently invisible to the eye; it works slowly, silently, undramatically; when diagnosed it often requires actions that are in conflict with deeply rooted social and religious values, life styles, and economic systems. In other words, the crisis is potentially lethal because it can only be met through levels of international cooperation unknown to world history" (Disch, 1970, xiii). J. F. Morrison explains the ecological basis of the problems confronting mankind thus: "Man's attempts to use and modify the gifts of nature to his own advantage have always had an effect on the natural environment—often a significant one—but it has only been in the last few decades that the level of our scientific discovery, machine technology, and organizational skills have made it possible on a world-wide basis for our population level and consequent demands on the environment for energy, water, air and other raw materials, open space, and waste disposal to grow beyond the capacity of our environment to provide for our needs (at least at current levels of knowledge, technology, and organization)" (Van Raay and Lugo, 1974, 180). J. McHale observes "It is only in the most recent, and brief, period of his tenure that man has developed in sufficient numbers, and acquired enough power, to become one of the most potentially dangerous organisms that the planet has ever hosted" (Miller, 1975, 5)

Many writers urge that ecology makes it evident that mankind faces a critical situation, indeed, "an ecological crisis." Thus we find R. A. Falk stating: "This book is about survival, its desirable modes and its possibilities. More specifically, the subject is political order on a global scale. . . . The human race has between ten and one hundred years left to fashion a response to a situation of growing danger" (Falk, 1971, 1). Boughey observes: "Within the past twenty years it has become apparent that we have produced too many people, too many pollutants, too much waste, too many poisons, too much stress. At the same time we have too little food, energy, shelter, education, health and understanding. We are squandering our global resources of fossil fuels, mineral ores, productive lands, wildlife, air, water, landscape, wilderness, and biotic diversity. Disaster looms on every horizon, both for our own population and for the ecosystems we occupy" (Boughey, 1975, 2).

Different writers stress different elements in what they characterize as "the ecological crisis": resource depletion, population size and growth, pollution, the endangering of species and loss of wilderness. But they agree in the fact and importance of the interconnectedness of the various elements and factors. The Ehrlichs stress population size and growth: "The explosive growth of the human population is the most significant terrestial event of the past million millenia. . . . Mankind itself may stand on the brink of extinction; in its death throes it could take with it most of the other passengers of Spaceship Earth. No geological event in a billion years . . . has posed a threat to terrestial life comparable to that of human overpopulation" (Ehrlich and Ehrlich, 1970, 1). Rachel Carson expressed alarm at pollution, more especially at the new forms of pollution caused by pesticides, arguing: "The most alarming of all man's assaults upon the environment is the contamination of air, earth, rivers, and sea with dangerous and even lethal materials. The pollution is for the most part irrecoverable; the chain of evil it initiates not only in the world that must support life, but in living tissues is for the most part irreversible" (Carson, 1962, 6).

The tenor of a great bulk of the writing relating to the relevance of ecology and to the threat of an ecological crisis is accurately reported by J. Maddox, who, while accepting the importance of ecology, expresses a view opposite to that expressed by many of the previously cited writers, namely that there is in fact no crisis, no problem or set of problems, with which man cannot adequately cope:

In the past decade, since the publication of Miss Rachel Carson's *Silent Spring,* the people of North America and, to a lesser extent, Western Europe have been assailed by prophecies of calamity. To some, population growth is the most immediate threat. Others make more of pollution of particular kinds, the risk that the world will run out of food or natural resources or even the possibility that economic growth and the prosperity it brings spell danger for the human race. And there is talk of potentially horrific uses of genetic engineering and even of the possibility that the temper of modern science may undermine the structure of modern society. But, although these prophecies are founded in science, they are at best pseudo-science. The most common error is to suppose that the worst will always happen. And, to the extent that they are based on assumptions as to how people will behave, they ignore the ways in which social institutions and humane aspirations can conspire to solve the most daunting problems. [Maddox, 1972, Preface]

Much of the concern expressed by ecological moralists is for human survival and the quality of life human beings will enjoy if they survive. A moral concern also is expressed by many, however, for the well-being of the earth's ecosystem, and its living members and inanimate components. Some theorists argue that the ecological crisis has arrived, that today the world is overpopulated and/or overexploited and/or irreversibly polluted; but most who write of "the ecological crisis" see it as being still in the future and preventable if we immediately take the appropriate action based on a full and proper understanding of the facts and laws of ecology, using our human intelligence and technology on the basis of such an understanding of the truths of ecology. By reference to predictions that unless we mend our ways in our treatment of our environment, disaster will overtake mankind and the earth, ecologically based moral and political measures are generally proposed as essential steps toward solutions to "the crisis."

As this is a philosophical, not a scientific, work, no attempt will be made to marshal evidence for or against the many diverse and often conflicting claims made by specialist scientists within and outside the boundaries of their scientific competence. But because, as will become evident in Parts II and III, so much, morally and politically, hangs on the soundness of the predictions, some attention must be given here to the bases of the predictions of a crisis; the problems encountered in attempting to ground soundly such a complex, cosmic prediction or set of predictions will briefly be noted in this part. Brief attention will also be given to the kinds of moral and political consequences that may result from ignoring a true and correct prediction of a crisis or from acting on the basis of a false, incorrect forecast that a crisis is imminent.

While many of the important, philosophically interesting issues of ecological ethics and politics arise on the basis of predictions of a crisis, the problems that are a matter of our concern are not confined to those that would arise only if a crisis were imminent. Hence, this work will be concerned with a considerable range of issues and problems.

In Part II the first major issue to be explored concerns the basis for preserving plant and animal species, wilderness and natural phenomena, forests, seas, lakes, and mountains. In this context, various moral philosophers increasingly urge that traditional ethical theories, because they are human-centered or heavily human-oriented, are inadequate as bases for explaining our duties and rights concerning nature and its constituents. Hence, some ecological moral philosophers argue that a new theory of valuation (that is, a new meta-ethic that does not explain moral valuation in terms of human attitudes, preferences, desires, and the like, as do the presently fashionable noncognitivist and naturalist ethical theories) is needed, one that explains valuation completely independent of man, his attitudes, wants, needs. These moral philosophers, together with other less radical ecological ethicists, insist also that a new normative ethic must be developed, one that acknowledges either the intrinsic value of nature and its constituents, species, wilderness, and forests, and/or the existence of intrinsic obligations to respect nature and its constituents, whether or not they possess intrinsic value. These ethical issues would arise even were there no threat of a crisis. Our duties in respect of nature, the preservation of species, wilderness, and natural phenomena, relate both to the case for retaining actual and possible future resources as renewable resources, and also to the question of whether nature and its constituents should morally be respected for their own sakes. The occurrence of a crisis due to resource depletion, overpopulation, and/or a worsening of pollution may well heighten the dangers to nature, but there would still be threats to nature and its constituents even were no such crisis to be likely to eventuate.

Another issue of moral importance to which the prediction of an ecological crisis is only marginally relevant is that of animal moral rights. Yet the possession of moral rights by some or all animals, and the recognition by mankind of such rights, is of ecological importance. An increasing number of philosophers—possibly even an exponentially increasing number—is arguing that certain animals possess moral rights. Many of these same philosophers see the recognition of and respect for animal rights as being of the greatest ecological importance by making clearer and more determinate our ecological duties. By contrast, other ethicists contend that, without further ecological ethical innovation, the acknowledgment of moral rights of sentient animals would simply result in an enlargement of the class of beings accorded the right to ravage nature. In Part II it will be argued that animals do not possess moral rights, and that this is fortunate if we are concerned about the environment and our duties in respect of it.

Predictions of an ecological crisis are most relevant to the moral and political concern about resource depletion (the using up of nonrenewable resources), the knowledge provided by science and the powers provided by technology, the control of pollution and of population growth. Those who perceive the ecological crisis arising wholly or in part from depletion of nonrenewable resources discern important moral issues in the conflict between the use of such resources today and our duties not to leave future generations without what is essential for their well-being, even though the use of resources today could alleviate the unfortunate condition of at least two billion persons, while the needs and numbers of future generations can only be a matter of conjecture. Others who see population growth leading to overpopulation and to an ecological crisis resulting therefrom, one that will involve resource depletion and a worsening of pollution, see the imminence of the crisis as central to the moral issues relating to man and his responsibilities in respect of nature. Science and technology are seen by various writers as being likely to facilitate access to new reserves of resources; to create new resources by transforming presently useless raw materials; to lessen demands on various resources, including energy resources; to provide the means of greatly increasing food production; and to make it easier, morally more acceptable, and practicable to check population growth by means of improved, more reliable methods of birth control. These issues call for moral discussion and resolution in terms of recognized ethical values, respect for persons and the rights of persons, justice, honesty, and the securing of the intrinsically valuable. Morally acceptable solutions need to be devised that rest on respect for persons and for such basic human rights as those to moral autonomy and integrity, liberty, the opportunity to be self-developing and to attain well-being, as well as to those equalities that are dictated by justice.

In Part III the ecologically based and ecologically relevant political issues are examined. Like the comparable moral issues, the ecologically based political issues are not confined to those arising from the possibility of an ecological crisis. Nonetheless, the reality of a threat of such a crisis would enhance the need to find some far-reaching, consequential solutions, not least the development of a powerful world political authority that is ecologically aware and crisis-conscious. Further, because none of the issues of ecological moral concern—preservation of nature and its constituents, the recognition of any moral rights animals might possess, the conservation of resources, renewable and nonrenewable, the proper development of science and use of the powers technology provides, the checking of pollution, and the prevention of overpopulation—is such that it can be resolved satisfactorily by voluntary action by individual persons, political action and political solutions are required.

The preservation of species, wilderness, and natural phenomena that is dictated by concern to retain actual or possible renewable resources for man's benefit or to preserve the intrinsically valuable creates no problem of political principle. States may legitimately use their powers and authority to such ends. The problems that need to be faced here are purely practical ones, relating to the devising of effective national and international machinery to achieve the desired goals. On the other hand, the preservation of species, wilderness, natural phenomena that do not possess intrinsic value—not for man's sake but for their own sakes, by states and the international political community—presents very real problems of principle concerning the possibility of justifying such political action, since it must necessarily involve the curtailment of the freedom of persons, their self-development, and their enjoyment of well-being, by means of criminal, taxation, and various other laws.

Were some or all animals successfully to be shown to possess moral rights, the resultant political implications would be considerable, both in respect of the political action that would be necessary to secure respect for them and because of the adverse ecological consequences that could follow upon such recognition of animals' rights. This could well make more probable or worsen the ecological crisis of which so many writers warn.

The politics of conservation of resources raises questions concerning both the goal to be aimed at and the methods to be employed. Problems here relate to our uncertainty concerning how much conservation of resources (and what kinds) is morally and politically desirable in terms of the danger of a crisis. Too little or the wrong kinds could be useless. More than is necessary would itself be wasteful, while both would involve unnecessary and undesirable violation of basic human rights, not least the rights to liberty, self-development, and to those equalities dictated by justice. Should it be necessary to opt for the goal of a resource-frugal world society, problems will need to be resolved concerning the appropriate political machinery to be used to achieve that goal as well as about how best to minimise the costs involved in terms of restrictions on persons in the exercise of their liberties.

The political issues relating to pollution are those of gaining acceptance of those political institutions and measures within states and within the world political community that are necessary to control pollution. They are measures readily justifiable in terms of most philosophies of the state including liberalism, even though their acceptance and use would transform liberalism from a seemingly negative, relatively noninterventionist theory of the role of the state to one that provides support for considerable interferences with the individual's liberty.

Population growth and the prevention of overpopulation raise many important issues of principle concerning the morally and politically acceptable measures required for effective control. That the goals of family planning and of preventing overpopulation are quite distinct, that the former rests on the free decisions of the individuals concerned while the latter may admit of realization only by and through action by the state, including coercion, make it evident that the prevention of overpopulation, if such a threat were to occur, would necessitate a basic rethinking of the role and rights of states.

Since many writers maintain that, because of the gravity and imminence of an ecological crisis, the urgency of the need for drastic action involving draconian measures makes it essential that resort be had to a solution through totalitarianism, totalitarian states, and/or a totalitarian world political order, it is important that the relative merits of approaches to ecologically based and ecologically relevant problems through totalitarian and liberal democratic states and world political authorities be examined.

2

The Bases for "Ecological Crisis" Predictions

Projections of Present Trends into the Future, and Predictions Based on the Finitude of the Earth

Many forecasts of an ecological crisis are based on nothing more than the projection of present trends into the future, with an emphasis on the exponential rates of increase in resource use and depletion, population growth, and pollution. Consider here the predictions of *The Limits to Growth*, its precursors and successors, and the numerous course textbooks influenced and inspired by such writings.

Many who forecast an ecological crisis stress the finitude of the earth—consider the popular analogies drawn between the earth and a spaceship, lifeboat, or island—this being used to suggest that because the resources of the earth are finite, they are in danger of being used up, of being insufficient to support future populations. The analogies are designed to stress the finitude of the earth, yet the earth, although finite, is nonetheless immense; it also has a vast, continuing, inexhaustible (by man) input of solar and gravitational energy waiting to be fully harnessed by our technological inventiveness, to supplement or replace the finite supplies of fossil and nuclear fuels. Other more thoughtful, better-based predictions are derived from analyses of trends and of those factors that bear on their continuation or modification in the future, for example, in respect of population trends, food production, resource use, technological developments, and the like.

What is not to be found, and this because of the limitations of human knowledge regarding the vast range of relevant causal factors, is a carefully documented set of predictions based on a causal analysis of the many factors that influence and will influence population size;

population growth; improvements in food production and in the use of the food that is produced; probable depletion rates of nonrenewable resources, renewable resources that are rendered nonrenewable and no longer resources, and raw materials generally, based on analyses of factors relating to their use, the effects of reducing waste, and the probability or otherwise of there being a reduction in waste in production and consumption. Further, while it is known that technological advances will be made by way of inventing new resources, creating new uses for raw materials, exploiting solar and other natural energy sources, improving food production, and the like, the limitations of human knowledge are such that we can only guess as to the kinds of advances that will be made in the future. Great though human imaginative powers are, we are unlikely even to envisage, let alone rationally predict, the revolutionary new discoveries that will be made and adopted, just as people a century ago were unable to imagine many of this century's major developments such as nuclear energy, plastics, synthetics, the jet, and genetic engineering. Science fiction writers such as Jules Verne envisaged some of the more remarkable achievements of the twentieth century, but although they envisaged some of these achievements they did not foresee the means and methods by which these achievements were to be realized. One of the more interesting, less inaccurate "prophecies" of a century and a half ago was made during the lifetime of Thomas Malthus by T. B. Macaulay, and one of such extreme optimism that, as Macaulay himself noted, he would have been deemed insane for making it. Macaulay observed:

If we were to prophesy that in the year 1930, a population of fifty million, better fed, clad, and lodged than the English of our time, will cover these islands, that Sussex and Huntingdonshire will be wealthier than the wealthiest parts of the West Riding of Yorkshire now are,—that cultivation, rich as that of a flower garden, will be carried up to the very tops of Ben Nevis and Helvellyn,—that machines constructed on principles yet undiscovered, will be in every house,—that there will be no highways but railroads, no travelling but by steam,—that our debt, vast as it seems to us, will appear to our grandchildren a trifling encumbrance, which might easily be paid off in a year or two,—many people would think us insane. [Macaulay, 1830, 76–77]

Predictions and Projections Distinguished

Projections and predictions are very different things and need to be distinguished carefully from each other. The Club of Rome in *The Limits to Growth* purports simply to be offering projections, not firm predictions, about what will happen in the future if man does not act drastically to modify present trends. It projects present trends as continuing and indicates what the outcome will be if they continue, stressing that it is offering only projections. "We shall emphasize just one more time that none of these computer outputs is a prediction.

We would not expect the real world to behave like the world model in any of the graphs we have shown, especially in the collapse modes" (Meadows et al., 1974, 142). However, even as early as the Introduction, predictions are made that certain developments *will* occur unless mankind acts consciously to prevent them. Thus it is stated: "1. If the present growth trends in world population, industrialization, pollution, food production, and resource depletion continue unchanged, the limits to growth on this planet will be reached sometime within the next one hundred years. . . . 2. It is possible to alter these growth trends and to establish a condition of ecological and economic stability that is sustainable far into the future. . . . 3. If the world's people decide to strive for this second outcome rather than the first, the sooner they begin working to attain it, the greater will be their chances of success" (ibid., 23–24; see also Commentary, 185–197). As the work develops, the writers switch more explicitly to the more ambitious positions of claiming to be predicting accurately what the future will hold *if* man does not expressly modify these trends; and this is how the book has been generally received and interpreted. The authors imply that they can accurately predict what the future will be, if man does not act drastically to modify present trends; that they can do this simply by uncritically projecting present trends into the future. How else can they rationally state, "We firmly believe that the warnings this book contains are amply justified, and that the aims and actions of our present civilization can only aggravate the problems of tomorrow" (ibid., 195). Many predictions about population growth—the more alarmist ones more commonly than other predictions—similarly start out as projections of present rates of growth, and then, without further argument, and without any attempt at justification, switch to being advanced as if they constitute reasonable predictions based on solid, sound inductive evidence.

The distinction between a prediction and a projection is of the utmost importance. There is nothing illegitimate about making projections of present trends, or double or half such, into the future, provided that those who make them are clear themselves and make it clear to their audience that they are not predicting, simply projecting, and saying that if present trends, or those hypothesized, prevail, this is what would happen, but that they do not know whether the present or the hypothesized trends will prevail. The authors of *Limits*, together with many others who forecast an ecological crisis on the basis of projections, attribute great value to what are simply projections, a value that only well-grounded predictions can have, as they are concerned to argue that their projections show that unless we mend our ways in our use of resources, our breeding, polluting, our management of our environment, disaster will occur. In fact, all they are entitled to assert is that if present trends continue unchanged into the future, where it is not known whether they will do so, this is what will occur.

The projections become predictions, but not well-grounded ones, if and only if it is assumed that present trends will continue, that natural developments, or human action not necessarily directed at that end, will not change them, that advances in technology as in food production, birth control, the extraction and recovery of minerals, the harnessing of natural, nonpolluting energy sources, and the like will not modify them. In fact, various factors have modified the trends relating to the different kinds of growth upon which such projection-predictions rest. What is of importance here is that for projections such as those of *Limits* to become predictions, they must be supplemented by important but very questionable assumptions. To assess the soundness of the predictions based on these assumptions, we should need to assess the soundness of each one of them. This cannot easily be done. With some assumptions, it is possible to give some sort of estimate of their probable truth or falsity. With others this is not possible. Thus, the fact that predictions such as those of *Limits* have been taken as seriously as they have by academics of the Western world suggests not so much that an ecological crisis is imminent but that there is a crisis of another kind, a crisis of academia. Were any of our political leaders to proceed to plan our nation's future on the basis of such blind assumptions, we should at best regard them as totally incompetent. More likely, we should regard them as mad. We know that, irrespective of our deliberate planning, present trends will not be maintained at their present levels during the next ten, twenty, fifty years; that there will be great technological developments, some resulting from existing developments and their inbuilt forces.

A point that is often made in criticisms of *Limits* illustrates this. *Limits* represents the limitations of nonrenewable resources as being essentially geophysical and seeks to take note of the possibility of new reserves being discovered by postulating the difference the finding of fivefold the present reserves would make. Quite apart from the fact that since the publication of *Limits* vast reserves of petroleum and other seemingly scarce nonrenewable resources have been found, so as to make the fivefold figure an evidently arbitrary "guess," this will not do for other reasons. The more basic criticism is that, although reserves are necessarily finite, since the earth is finite and hence if nonrenewable and if continued to be used, will eventually be used up, what count as resources and as reserves of most raw materials are so counted by reference to technology and economics. Given how little of the earth has been explored for those raw materials we call resources, how shallow a depth of the earth's crust has been penetrated, how little use has been made of the minerals in the oceans, then, if economic and technological difficulties are ignored, the reserves of most resources would be seen to be immense. The real limits to reserves of most resources are economic and technological. This is important. It means that we must seek to assess whether these limitations are such that they can be overcome, immediately or in the longer term. Physical limits cannot be overcome; they provide an absolute barrier. Economic and technological limits may be circumvented. A further point relates to the replaceability of resources by new raw materials in a way that makes it irrelevant what the size of the reserves of the obsolete resource is.

Parallel points may be made about population projections that come to be advanced as if they are soundly based predictions. Even before effective contraception, there were varying rates in population growth and decline. Demographers offer little guidance as to the causes of fluctuations in birth rates, although they can say a good deal about factors bearing on death rates. It is possible but unproven that economic and social changes—a higher standard of living, greater income, a transition from agriculture to urban, industrial life and work, and greater equality of status and power for women—may permanently lower birth rates. An unstated assumption of many who advance such views is that women, if free and independent, would choose to have fewer children than they now do in high-birth-rate countries. It is also possible but unproven that there is some connection between a falling death rate and a subsequent falling birth rate. The lack of soundly based knowledge together with the fact that contraception and other methods of birth control are becoming increasingly available and effective and more widely used make it evident that it is unrealistic and mischievous to treat uncritical projections of present trends as serious predictions. Yet one finds many writers, of whom R. A. Falk in *This Endangered Planet* is representative, citing such projections with approval, as if they lent serious support to the thesis that this planet is in danger of being overpopulated, and endangered by such overpopulation. Falk, citing Sax's projections that by 2171 Mexico alone could have a population of 40 billion, writes that if the present rates of increase were to continue, the world in 2105 would have a population of 36 billion and possibly be in danger of running out of oxygen. (Falk, 1971, 141–142).

In brief, most of the predictions of an ecological crisis lack a sound scientific basis. They rest not at all on the findings of the science of ecology and are ecological only in that the ecological interactions of the various factors contributing to the crisis are stressed by those who make such a prediction.

Problems in Predicting the Future: Scientific, Inductive Criteria of Sound Predictions

The complexity of the relevant factors, our very limited data, the limited knowledge as to what constitute relevant factors and data (in particular, what are and what will be the relevant causal forces and processes at work), the very limited time span to which our knowledge and data relate, all contribute to making predictions about man's future, whether it be the near future of 50 years hence or the more distant futures of 100, 200, 500, or 1,000 years, uncertain and highly fallible. The more distant the future about which we are seeking to make the predictions, the more uncertain and fallible they are. Yet even the future of 500 or 1,000 years is a near future in terms of man's existence.

The difficulties inherent in making accurate predictions about future developments of society may be demonstrated by considering the kinds of predictions which, on the evidence then available, might rationally have been made about today in 1932, 1882, 1832, 1782, 1482 or 982. What reasonable hope could forecasters have had in those years, on the basis of facts then available to them, of predicting the developments that have since occurred? Jevons in 1865 was concerned about the possible exhaustion of coal in Britain. Oil was not seen as a significant source of energy. The role and effects of the internal combustion engine, of aluminum, synthetics, plastics, and the pollution to which they give rise, could not possibly have been predicted. Equally, the development of and our dependence on pesticides used in food production, and on such life-saving drugs as the sulfa drugs and other antibiotics, could not have been anticipated even 100 years ago. Indeed, even predictions made 50 years ago about what raw materials would be used as resources, about population sizes, rates of growth, rates of economic growth, and the nature and extent of pollution could not have been other than inaccurate. Much of the impact of writings such as Rachel Carson's *Silent Spring* turned on their imparting new information about unanticipated developments that followed other developments accelerated by World War II and led to new, dangerous forms of pollution, which at best could only have been speculated about in 1932.

That many of the recent forecasts of crises and calamities that have been made regarding events in the present time have been seriously mistaken, and mistaken by way of erring on the side of confidently predicting disasters and crises that have not eventuated, must heighten doubts about the credence to be given to such general forecasts. Typical of this kind of mistaken, alarmist prediction was the forecast of W. and R. Paddock (in *Famine-1975!*) that the time of famines would be upon us full-scale by 1975, quoted with such approval by P. R. Ehrlich in 1968 in his article "World Population: A Battle Lost?" (Anderson, 1970). Similar mistaken near-contemporary forecasts include predictions concerning population growth in Europe and North America, food production levels, levels of material consumption and, in reference to the past decade, rates of use of petroleum and rates of economic growth.

The Relevance and Irrelevance of the Computer

A computer allows us to make predictions from known data more quickly and accurately, to work out the complex interactions of various forces and factors. It in no way improves on the data that is fed into it. The Club of Rome made much of its use of the computer, and much of the impact of *The Limits to Growth* was due to the fact that people are impressed by computers as being more reliable than fallible men. However, among the more legitimate criticisms of *Limits* was "Garbage in, garbage out"—that is, a computer does not check data and cannot improve upon the data that is fed into it (Passell, Roberts, and Ross, 1972). Hence, if what is fed in is a guess or false assumption, the computer cannot make anything other than guesses or assumptions. Accuracy in respect of the complex, multifarious data is all-important when seeking to forecast the developments that will occur in the world at large during the next 50, 100, 200, or 500 years. The computer is of no value without such accurate, complete data.

Scientific, Inductive Criteria

No reputable scientist in any established science in areas other than predictions of world population growth, the effects of such growth, resource use and depletion, pollution growth—that is, areas relating to "the ecological crisis"—would dream of making predictions on the basis of an uncritical, unanalyzed acceptance of present trends. The first move any serious physical or social scientist will make is to analyze data in an effort to determine what causal factors are operating, the factors with which the trends are correlated, whether the trends are explicable in terms of presently established theories and, if so, whether such theories suggest that the trends are likely to continue unchanged. Where mere short-run trends are available, and where there are evidently multifarious causes at work producing the trends, it may well be impossible to make a prediction that has any significant probability value. The size and range of the sample are relevant to any probability value that such a prediction will have. Very few key predictions supporting the claim that we are faced by an ecological crisis measure up to such inductive, scientific standards.

Key Concepts Used in Predicting an Ecological Crisis

Predictions of an ecological crisis rest on claims that make use of *resource, nonrenewable resource, reserve, overpopulation, optimum population, and pollution* as key terms. They are often used as if they point to eternal, objective, Platonic Forms, instead of to things so characterized at least in part on the basis of economic, technological, and moral criteria. This greatly qualifies the predictions in which they figure or which depend on reference to them. Their use is in need of explanation and clarification, since some present usage is confused and equivocal.

Resources and Raw Materials

Apart from some very basic raw materials that are essential for physical, organic existence and life—things such as water, air, basic amounts and kinds of foods—and hence the elements of which they are composed, many so-called resources are socially, technologically, or economically relative to specific societies; thus, many of today's resources, including fossil fuels, uranium, and mercury among others, have not been and are unlikely to remain resources. Fossil fuels came to be resources only at a certain stage of development; before that they were no doubt seen as sources of natural pollution. More generally, throughout human history, technology has rendered to be resources what were not such, and equally, not to be used as resources, what had formerly been such. There are serious dangers in seeming to eternalize as Platonic Forms, as eternally resources, those raw materials that, in 1982, are seen to be and are used as resources. Today's waste material may be tomorrow's resource, and vice versa. Hence it is that much talk about using up essential resources presupposes a static stage of society and technology.

Nonrenewable Resources

Great importance is attached to the distinction between nonrenewable and renewable resources, as if some sort of intrinsic distinction is to be drawn between those resources which are used up in being used and those that are not so used up. In fact, this distinction too is essentially technologically relative with most resources. Among the so-called renewable resources are air and land; among the nonrenewable resources are fossil and nuclear fuels, metals such as mercury, zinc, iron, and aluminum. The thought is that the nonrenewable resources are used up in being used. If we use coal, mercury, and the like, less will be left for future use. However, if we look more closely we see that this is not a completely accurate account of the matter. Mercury, iron, and aluminum are not used up, lost to Spaceship Earth. They are simply dispersed and at present cannot fully be recovered. The *cannot* here is an economic and technological one. With fossil fuels, the situation superficially seems different. Coal and petroleum, in being used up, are destroyed as coal and petroleum. Nature may, over a long period of time, replace them, but meanwhile they are lost to the earth. In fact, however, their constituents are not lost. Even if possessed of the scientific and technical knowledge necessary to synthesize and reconstitute them, we should be barred from doing so by the costs in terms of the money and energy required. Nuclear fuels seem to be genuinely nonrenewable resources. Relevant to all this is the fact that the earth has a virtually inexhaustible supply of natural energy waiting to be harnessed by technology. That renewable resources—land, seas, forests—even with the most careful use and management may be rendered in part or in whole nonrenewable further blurs the distinction.

Science and Technology

The distinction between renewable and nonrenewable resources is also misleading in its exclusion of science and technology from the class, *resources*. They are important as sources of resources, and hence, as themselves, resources of a kind.

Overpopulation

Talk about the danger of world overpopulation goes hand in hand with suggestions that there is an optimum population for all time at which mankind ought to aim. *Overpopulation* is often used to refer to a state of affairs in which the population exceeds that for which there are adequate food and other basic essentials. *Optimum population* is commonly explained in terms of a population size that is compatible with life at what is considered a desirable standard of living; different writers settle for different standards as desirable. In this work, it is proposed to use *overpopulation* to refer to population sizes that exceed a country's or the world's capacity to provide adequately for the enjoyment of the basic human rights of all who are born into that country or into the world. Individual countries may suffer from overpopulation yet the earth not be overpopulated. The earth is overpopulated when it as a whole world community lacks the resources that are essential to provide all who are born, even when the resources are distributed in the fairest, most beneficial manner possible, that which is necessary for the enjoyment of their natural rights as persons. It is not possible to indicate, as do many writers, what constitutes overpopulation for all time, as this will depend on the state of technology, more especially in food production, the development of new, more productive varieties of crops and animals, on variations and changes in the earth's climate, and innumerable other factors.

Optimum World Population

Most writers who argue for controls to achieve the optimum population size seem to have utilitarian—hedonistic or ideal utilitarian—criteria in mind. Yet there are obvious reasons for rejecting such criteria, reasons noted even by utilitarians, which have led many to compromise their utilitarianism. Few theorists are prepared to argue that the optimum population is that which, irrespective of the total amount of suffering, has the greatest possible balance of pleasure. Rather, they argue for that which has the highest average pleasure or the attainment of a certain quality of life by most. Clearly, any defensible view of optimum population must be one that does not admit of suffering and misery on a large scale. To be morally acceptable, the concept must be explained in terms of a population that is compatible with all the basic human rights of those who are born being capable of being respected and that allows for justice and full respect for persons as persons and the enjoyment by many of goods beyond that to which all individuals have a right. In defining what constitutes the optimum population for any given time, different views reflect different emphases on various goods and their importance; thus, it is possible for different theorists to argue rationally for different levels of goods, different qualities of life. As with overpopulation, there is no optimum population size that holds for all time. The very misleading analogies between the earth and a spaceship, lifeboat, or island with fixed, limited resources help to foster such simplistic thinking.

The reasonable goal in respect of population is twofold: to determine what, given the facts as we know them, would be the likely range of population sizes from a defensible notion of *optimum* to a maximum that falls short of overpopulation; then to seek to contain population growth within that range until we have further information on the basis of which the goal may be enlarged or reduced. Relevant here is the fact that if there should be clear evidence of a danger of overpopulation, then full account needs to be taken of and allowance made for the momentum of population growth, and the time taken to stabilize a population size without social disruption, hardship, and great cost. Also relevant is the fact that total population size is not the only important issue in respect of population growth. A·very rapid rate of growth may cause great hardship, difficulties, and disharmony, because of society's inability to provide the necessary housing, health care, and educational facilities quickly. Many present world population problems appear to result from the unjust distribution of resources, combined with the rapid rates of growth of populations in countries most adversely affected by the world's distribution of wealth and the trading arrangements, barriers, and tariffs that help to maintain this distribution.

Pollution

Pollution, involving as it does at worst a threat to human life and health and at least harm, nuisance, and inconvenience to existing and sometimes future generations of persons, to animal and plant life, to wilderness, and to aesthetically valuable artifacts, raises important moral and political issues of respect for persons, of justice, prevention of evil, promotion of good, of honesty and fair dealing. Yet there are problems in defining *pollution*.

First, there is the problem of whether the concept is to be restricted to humanly caused pollution or whether it also encompasses naturally caused pollution—volcanic ash, dust, animal excreta, natural oil gushers, marsh gases, and the like. Any decision here must be stipulative, as both usages have wide acceptance. That we think of pollution as something it is morally wrong to bring about, as that for which those who cause it should be punished and made to pay reparations, provides point for adopting the narrower usage. So too does the fact that much of nature's "pollution" is admired as beautiful—coral reefs, volcanoes—or valued for its utility—as are phosphate islands. Yet pollution is generally something we wish to reduce or prevent, and this is usually true of naturally caused pollution as well as that caused by man.

Second, there is the subjective use of *pollution* to refer to what is simply unwanted by us, and the objective use to refer to what is, objectively, a nuisance, an inconvenience, something harmful or dangerous to us. Again both usages are to be encountered. Insofar as we see the polluter as one who acts unjustly, dishonestly, as one who harms others, who does not pay for the harm he does, who externalizes the costs of pollution, who is deserving of punishment and who ought therefore to be subject to the sanctions of the law, we are using *pollution* in the objective sense. This would seem to be the usage that morally and politically is most useful to adopt. This is not to suggest that the sustained subjection to that which we dislike may not become harmful to us. When it does, then it becomes pollution in the objective sense.

Attempts to define *pollution* are complicated by the fact that not all that harms us is pollution. *Pollution* as a concept overlaps various other concepts such as *annoyances, nuisances, poisonings, harmings;* what is rightly characterized as pollution may be any one of these things, but they need not be and are not always forms of pollution. The difficulties encountered by representative definitions bring out the nature of the problem. Miller defines *pollution:* "A *pollutant* is any material or set of conditions that creates a stress on or unfavourable alteration of an individual organism, population, community, or ecosystem beyond that found in normal environmental conditions. The range of tolerance to stress varies considerably with the type of organism and the type of pollutant, and determining whether an effect is 'unfavourable' may be a difficult and highly subjective process" (Miller, 1975, 267). This definition is unsound because not all things that have the effects pointed to are pollutants. Sustained artillery fire, a scarecrow, or well-directed pesticides that kill only the intended pests are not normally characterized as pollutants. There is a certain arbitrariness in the way we mark off more serious harms such as those that occur in warfare, and pollution. With pesticides, intention is relevant to our characterization of the situation. Pollution is typically but not always thought of as causing unwanted, adverse changes in the ecosystem, organism, or the environment; the more *adverse unwanted or unfavorable* are explained purely subjectively, the weaker is the moral case for condemning pollution.

Another definitional approach is in terms of noting the contrast between *pure* (pure water, pure alcohol, pure gold, clean, pure air) and the *impure,* where the impure has foreign substances in it. This approach is possible only in relation to a limited number of pollutants. If adopted, it would radically revise the ordinary concept, since well-directed uses of pesticides or the addition of chlorine to drinking water and preservatives to meat are not seen as causing pollution. J. Passmore seeks to explain pollution so as to take note of the fact that almost anything can become a pollutant when it occurs in an unwanted place. Hence he writes: "First, a definition. The classical definition of dirt is 'matter in the wrong place'; pollution is simply the process of putting matter in such a place in quantities that are too large. Or, more broadly, matter and physical processes—allowing for our present purposes, that everyday distinction—since radiation and noise commonly count as pollutants" (Passmore, 1974, 45). This account correctly brings out that most substances and processes can become pollutants, but it is incomplete in that it fails to mark off pollutants from other evils. Further, it needs completion by giving a reference to *too much. Too much* is commonly but not always explained by reference to human interests and concerns.

In brief, a completely satisfactory definition of *pollution* seems not to be possible. It can generally be characterized as matter, processes, or states of affairs that affect the environment and are caused by human actions that adversely affect human well-being and/or interests, and/or those of valued organisms or phenomena where what causes the adverse effect is not intended to have that effect, or, where intended, is nonetheless unwanted by man, and where some other description such as "poisoned," or "wounded" is not more apt than one in terms of harms caused by pollutants.

It is clearly important that the nature of these concepts and the relativity of the former concepts be properly understood when formulating, assessing, and using projections and predictions of crises that make use of these concepts.

3

The Politics of Forecasting and the Political Implications of Predictions

Predictions about the future, the obvious ones of the danger of an ecological crisis, such as those of the Club of Rome, the Paddocks, the Ehrlichs, G. Hardin, and the like, as well as the complacent optimistic ones implicit in many nonecologically based writings, have very important political implications, and they are commonly made with these implications in mind. Equally, they commonly have important political effects and often are made in the hope that they will have these effects. This is acknowledged in many of the writings. Different political groups within states, as well as states themselves, are profoundly influenced by different kinds of predictions. We need to know which kind of prediction, if either, we should accept, which implications are those of which account should be taken; equally, if neither kind is soundly or scientifically based, if neither has any right to our attention, we need to know that too, or be seriously misled in our political deliberations, planning, and decisions.

The predictions of crises have obvious, far-reaching political implications and ramifications. First, the worse the future is painted, the more ready we shall become to ignore present problems of poverty and injustices within the world community in order to concentrate all our efforts on survival, possibly even, as Hardin and other U.S. ecological moralists suggest, on our own and our community's survival. After all, if we as a country, or species, cannot survive without drastic measures, we are wasting our time and efforts aspiring to improve the lot of those who are doomed. Second, predictions of crises are almost invariably associated with a demand for a no-growth (wrongly confused with a resource-frugal) economy. This questioning and rejection of the desirability of economic growth (properly measured growth, not simply growth in the misleading GNP sense) has long been a feature of predictions of crises. Clearly, we need to have very good reasons for closing the avenues to innovation, mobility, and the elimination of unjust inequalities that a growth society opens up to us, and for opting for the severe restrictions on liberty a no-growth society necessitates. Third, if the predictions of crises are sound, they imply that we must discourage the poorer, underdeveloped countries of the world from aspiring to attain not simply Western levels of living, but even levels substantially higher than they now experience.

This point is often put in the form that today's malthusianism is in the interests of the rich countries, that it is favorable to securing the interests of the privileged minority today. Hardin in "Lifeboat Ethics: The Case Against Helping the Poor" (1977) and D. C. Pirages and P. R. Ehrlich in their advocacy of U.S. ecological imperialism, among many others, make this very clear. Many Third World thinkers are aware of this and reject both the predictions of crises and the evidence on the basis of which they are made. Yet clearly, if there is a danger of a world crisis, and if the underdeveloped countries were to use resources at the rate at which developed countries now use them, the crisis would be hastened. Even to use only fivefold, not twentyfold, their present resources would seriously aggravate the situation. As is acknowledged implicitly or explicitly by many theorists, it is easier to check growth in the use of resources by underdeveloped countries than to reduce total resource use in the developed countries. Thus, an acceptance of predictions of crisis must inevitably lead to a change of attitudes toward underdeveloped countries.

Further, since the underdeveloped countries are also countries in which population growth is highest, such predictions must also, if accepted, lead to a concern, even a demand, that they reduce their population growth and even bring it quickly to a halt. There will be an increasing lack of sympathy with their plight if they do not respond to the demand that they cut the rate of population growth, even though tariff barriers, trade practices, and restrictions on immigration do much to contribute to their problems.

It is therefore vital that, before such predictions are accepted and acted upon, their claims to reliability be carefully assessed. They appear to rest on very inadequate, flimsy bases. Yet they lead to the boldest of political demands and proposals, the Paddocks writing off India as beyond salvation (Anderson, 1970, 45; see also Paddock and Paddock, 1967), while Ehrlich urges that "we should (1) Announce that we will no longer ship food to countries where dispassionate analysis indicates that food-production unbalance is hopeless. (2) Announce that we will no longer give aid to any country with an increasing population until that country convinces us that it is doing everything within its power to limit its population" (Anderson, 1970, 21).

Although some environmentalists such as R. Routley and V. Routley believe that solutions to environmental problems can be found in forming anarchical communities, most theorists see the solutions to be available only in terms of increasing statism, increased centralized planning and control, and increased restrictions on individual liberty, to control the economy, check resource use, population growth, pollution, damage to the environment, and to plan urban and other development. Some see a fully authoritarian, totalitarian state as being necessary. The implication of most discussions is that a strong world political authority is essential. This means forgoing both the enjoyment of many rights and liberties and opportunities to rectify grave injustices. If the predictions are sound, this may well be what we shall have to do. (This will be discussed in Part III.) But again, before we opt to make such sacrifices, we need to be rationally confident that the forecasts of crisis are soundly based.

Political bias may enter into predictions both in the collection of the data and in its interpretation. In writings about the ecological crisis one constantly encounters charges and counter charges that economic forecasts, population predictions, estimates of resources, future food production, and the like are determined by whether the forecaster is a supporter of a growth economy. Karl Marx condemned Malthus and his doomster forecasts as politically biased and motivated. Such charges and countercharges have been commonplace ever since. Marx wrote "The hatred of the English working class against Malthus—'the mountebank parson' as Cobbett rudely calls him—is therefore entirely justified. The people were right in sensing instinctively that they were confronted not with *a man of science* but a *bought advocate*, a pleader on behalf of their enemies, a shameless sycophant of the ruling classes" (Marx, 1861–63, 123).

Although the ecological crisis movement advances radical proposals regarding resource use, population control, pollution, and protection of the environment, it is essentially conservative in the sense of seeking to sustain the status quo. Its arguments in terms of a threatened crisis are well-directed toward that end and are in the best traditions of theoretical and practical politics, that of painting the alternatives as bleakly as possible. (Compare with, and compare Hobbes's and Locke's accounts of the state of nature and the conclusions about our obligation to obey the law they sought to derive therefrom; compare the ecological crisis style of argument with that of post–World War II nuclear disarmers whose slogan was "Better Red Than Dead.") Man clearly has the capacity to do great harm to himself and the earth he inhabits. However, the concern of Part I has been that predictions of an ecological crisis that emanate from concern about resource depletion, overpopulation, excessive pollution, or damage to the environment be critically and rationally assessed before they are accepted as a basis for moral and political action, that important values not be given up lightly nor the enjoyment of basic rights forgone without there being good and adequate reasons for making these sacrifices.

PART II

Ecological Ethics

4

The Relevance of Ecology to Ethics and Morality

There are three kinds of views concerning ecology's relevance to ethics and morality. One general view is that ecology and the awareness of nature, its constituents and their interdependence that it brings, necessitate the development of a fundamentally new ethic, a new, non-human-centered, nature-centered morality. A second and less radical view is that the findings of ecology make it necessary to develop a new ethic, but one that is new in a less radical sense, namely in the sense that it is ecologically orientated, ecologically aware, a normative ethic that fully acknowledges environment values and environmental duties. The third kind of view, and the one to be set out and defended in this work, is that the findings of ecology do not necessitate a basic revolution in ethics but simply a more informed, more accurate thinking out of our moral obligations and moral rights.

The view that a new ecological ethic that explains valuation as not man-dependent, one that fully and adequately indicates the values and the valuable in nature, that brings out that natural phenomena and not simply man demand respect and which explains man's duties regarding himself and his environment (noting that his is a very modest place in the moral scheme of things), is one that is more commonly alluded to than set out and defended in a developed, worked-out form. Indeed, while many write as if such an ethic exists and is the true ethic, it would seem that no moral philosopher has yet succeeded in either stating or defending it in a systematic way. Nonetheless, it is a view explicitly espoused by R. Routley and V. Routley in various articles; it is also a view to which Aldo Leopold gave voice when he claimed that we need a land ethic, a conservation ethic, one that enlarges the membership of the moral community by including all the constituents of nature, and when he suggested that what is right, is right in so far as it tends to preserve the integrity and stability of the biotic community. To this end he wrote: "All ethics so far evolved rest on the single premise: that the individual is a member of a community of interdependent parts. . . . The land ethic simply enlarges the boundaries of the community to include soils, waters, plants, and animals, or collectively, the land" (Leopold, 1966, 219). And: "A thing is right when it tends to preserve the integrity, stability, and beauty of the biotic community. It is wrong when it tends otherwise" (*ibid.*, 240).

Leopold's approach to a specifically ecological ethic can be developed along the lines of an ethical holism, with nature being either personified or seen as an organism, with man viewed as an organ of the natural organism, and not necessarily as the most important of the organs. C. D. Stone, in *Should Trees Have Standing: Towards Legal Rights For Natural Objects,* suggests but does not develop another kind of organic view underlying another distinctively ecological ethic when he writes: "The problems we have to confront are increasingly the world-wide crises of a global organism: not pollution of a stream, but pollution of the atmosphere and of the ocean. Increasingly, the death that occupies each human's imagination is not his own but that of the entire life cycle of the planet earth, to which each of us is as but a cell to a body" (Stone, 1972, 53).

Typically, the drive to develop a new ethic arises, as it does with the Routleys, from a concern to explain the value of natural phenomena, species of plants and animals, forests, rivers, wilderness, and the like independently of human wants, needs, preferences, and valuations. The problem of that ecological ethics is that of devising a new meta-ethic, one that explains valuation, and hence values, without reference to human beings, their preferences, pro-attitudes, and the like, and a new normative ethic that explains both the value of natural phenomena prior to man's existence and after the extinction of the species *Homo sapiens* and why it is morally wrong for man to damage wilderness and endanger or render extinct natural species of plants and animals.

The second, less radical, approach is that of developing a new normative ethic by way of modifying and correcting the traditional ethical theories so as to acknowledge new, specifically ecological values and duties. Not surprisingly, this approach is often run together and confused with the first, more radical, approach so that it is not always easy to determine in which sense a specific writer is intending to insist on the need for a new ethic. (For a useful, relevant discussion, see Blackstone, 1979.) Consider such statements: "Ecological ethics queries whether we ought again to universalize, recognizing the intrinsic value of every ecobiotic component" (Rolston, 1974–75, 101). And: "The conscience of ecology must teach men that there is no natural right to exterminate a form of life; that one is not entitled to desecrate earth, air, water, or space merely because he happens to own, control, or occupy some portion of it; and that the fact of 'legality' in a human court cannot remove ecological crimes from having planetary implications for all mankind" (Disch, 1970, 17–18).

The two distinct claims, the one that certain animals (usually identified as sentient animals) possess moral rights, and the other that living organisms and even natural phenomena should be accorded legal rights, are associated, on occasion, sometimes with the more radical claim that a new distinctively ecological ethic is possible, necessary, and such as can be formulated, sometimes with the less radical view that traditional ethics simply needs to be supplemented by reference to ecological values, among which some list animal moral rights, others those features of natural phenomena and living organisms which provide a basis for according them legal rights.

The third view, that which will be defended here, is that there is no need for a specifically ecological ethic to explain our obligations toward nature, that our moral rights and duties in respect of nature and its constituents can satisfactorily be explained in terms of an ethical theory of the kind developed by W. D. Ross in *The Right and The Good* and *Foundations of Ethics,* supplemented by an account of moral rights of persons: specifically, an ethic of respecting persons, of justice, promotion of good, of honesty, and of respect for basic human rights such as life, health, respect for persons, bodily integrity, moral autonomy and integrity, knowledge, and self-development (including education). In terms of this view, ecology bears on ethics and morality in that it brings out the far-reaching, extremely important effects of man's actions, that much that seemed simply to "happen"—extinction of species, depletion of resources, pollution, overrapid growth of population, undesirable, harmful, dangerous, and damaging uses of technology and science—is due to human actions that are controllable, preventable, by man and by men, and hence such that men can be held accountable for what occurs. Ecology brings out that man, often acting from the best of motives, often, however, simply from shortsighted self-interest without regard for others living today and for those yet to be born, brings about very damaging and often irreversible changes in the environment, changes such as the extinction of plant and animal species, destruction of wilderness and valuable natural phenomena such as forests, lakes, rivers, seas. He depletes natural resources, including land, plant and animal resources, fossil fuels, varieties of whale, fish, and other food resources, and he disperses many resources into the environment, making them presently unavailable as resources. He does harm by his disposal, and nondisposal, of waste products of his living, his manufacturing processes, and energy production. Many reproduce at a rate at which *their* environment cannot cope, so that damage is done to it, and, at the same time, those who are born are ill-fed, ill-clothed, ill-sheltered, ill-educated. Science and technology have reduced death rates and have given us greater power to control birth rates. They have conferred on us vast powers that can be used for both good and evil purposes, to kill, destroy, create new varieties and kinds of organisms, to increase food production, to control disease, and also to destroy the equilibrium of the world ecosystem in a way harmful to man.

Moralists concerned with the environment have pressed the need for a basic rethinking of the nature of our moral obligations (and of our moral rights) in the light of the knowledge provided by ecology on the basis of personal, social, and species prudence, as well as on general moral grounds in terms of hitherto unrecognized and neglected duties in respect of other people, people now living and persons yet to be born, those of the Third World and those of future generations, and also in respect of preservation of natural species, wilderness, and valuable natural phenomena. Hence we find ecological moralists who adopt this third approach, writing to the effect that concern for our duties entails concern for our environment and the ecosystems it contains. R. Dubos writes: "Were it only for selfish reasons, therefore, we must maintain variety and harmony in nature. . . . The cult of wilderness is not a luxury; it is a necessity for the protection of humanized nature and for the preservation of mental health." "Man can manipulate nature to his best interests only if he first loves her for her own sake" (Dubos, 1973, 166–167 and 45). And Pirages and Ehrlich write: "At this very moment, mankind is escalating its attack on the life-support systems of the Earth at a rate that will double human impact every fourteen years. . . . More optimistically, if those systems can take fifty times the level of abuse of 1970 before they disintegrate, then industrial society cannot survive past the middle of the next century, unless the rate of human attack on nature is slowed" (Pirages and Ehrlich, 1974, 2). And J. Passmore: "What it needs, for the most part, is not so much a 'new ethic' as a more general adherence to a perfectly familiar ethic. For the major sources of our ecological disasters—apart from ignorance—are greed and short-sightedness, which amount to much the same thing. . . . There is no novelty in the view that greed is evil; no need for a new ethic to tell us as much" (Passmore, 1974, 187).

The nature of the ecological ethic that is needed to explain and underpin our moral obligations will be argued for by way of a consideration of the nature of our duties in several areas: preservation of species, wilderness, and valuable natural phenomena; whether animals, some or all, possess moral rights; what are our obligations in respect of the use and depletion of resources, bearing in mind the rights and needs of those now living and those of future generations; our duties regarding pollution and of polluters; and our rights and obligations in connection with reproduction and population growth.

It will be argued that our duties and rights can satisfactorily be explained and grounded in an ethic that consists of principles of *prima facie* obligation to respect persons, be just and honest and promote good, and of basic moral rights and various conditional, derivative moral rights, where the basic moral rights are the rights to life, health, bodily integrity, respect as persons, moral autonomy and integrity, self-development (including education), knowledge and true belief, and justice, where these principles and rights are interpreted and applied in the light of the information provided by the sciences, including the science of ecology.

5

The Preservation of Nature

The issues to be considered in this chapter are those that most typically raise the question of whether there is a need for a radically new ecological ethic in terms of which it is possible to exhibit and explain values inherent in nature, and in terms of which it is possible to derive and determine man's duties. The issues relate to man's obligations not to endanger or render extinct species and varieties of plants and animals, not to encroach on or destroy wilderness areas, and not to damage natural phenomena, rivers, seas, lakes, mountains, valleys. Those who argue for a new ethic claim that the moral importance of preserving species and wilderness and respecting nature cannot adequately be accounted for in terms of existing ethical theories. Here it will be argued, against this contention, that while a very strong case for preservation of species and wilderness can be founded on traditional ethics, specifically on the ethic outlined above, there is no moral case for preservation when preservation of species and wilderness is contrary to respect for persons and for their rights, or when it is not dictated by promotion of good: that no successful case for a new ethic, a non-human-oriented ethic, nor for an ethic based on a holistic organicism where nature is viewed as an organism, nor on a pantheism or mystical view of nature, can be developed by way of argument from the nature of our duties in respect of preservation of species and wilderness. While it is clearly true that we hold the beautiful and mysterious in nature in awe and wonder, see much as precious, and view with moral shock and even moral outrage the willful destruction of a beautiful species, we equally, and rightly, deplore attempts to keep in a healthy, secure state such dangerous species as disease organisms, harmful, disease-carrying parasites, and the like when this is not dictated by concern for human well-being. Equally, we deplore the failure to develop fertile areas of wilderness when the failure to do so results in the deaths from starvation and the diseases associated with malnutrition of millions of persons.

The terminology followed here is that used by J. Passmore. Passmore explained preservation as consisting of the "attempt to maintain in their present condition such areas of the earth's surface as do not yet bear the obvious marks of man's handiwork and to protect from the risk of extinction those species of living things man has not yet destroyed" (Passmore, 1974, 101). This definition of preservation contains within it an account of wilderness of the kind most commonly used or assumed by writers who argue for the preservation of wilderness. *Wilderness* proper would seem to be that which is completely untouched and unaffected by man. So defined, there probably is little, if any, wilderness left—consider the far-reaching effects of man, including early man who fired the forests and cleared vast areas, and the extent of pollution, which penetrates such remote areas as the Arctic and Antarctic. We cannot now know what areas are genuinely wilderness, or even whether any really be such. Hence it is that those who express concern for preservation of wilderness mean by that expression what Passmore suggests: namely, those areas that seem to be relatively unaffected by humans. It is concerning man's responsibilities in regard to these areas that the case for preservation of wilderness is argued.

The findings of ecology considerably strengthen the case for preservation of species and wilderness—the two are interconnected—not by establishing that they possess intrinsic value, since ecology, as a science could not show this, but by showing that preservation is commonly dictated by conservation of resources, that much that at present seems to have neither intrinsic nor instrumental value really has, because of the interconnectedness of things of nature, great instrumental value, and much, with the advance of science and technology, will have instrumental value of various kinds in the future. In such ways, ecology shows that man needs to have very good reasons for not preserving species and wilderness. However, man can, and often does, have good and adequate reasons for endangering species and encroaching on wilderness, even though it is clearly true that he quite often wrongly, unnecessarily, unjustifiably endangers species, destroys wilderness, and in so acting brings about irreversible changes that are actually or potentially disadvantageous to man. In so arguing, the contentions of ecological ethicists such as Aldo Leopold, R. and V. Routley, and many others, are being rejected. Leopold argued that there is a need for a new ethic, a land ethic that would explain and justify our ethical intuitions; he wrote: "His [civilized man's] sense of right and wrong may be aroused quite as strongly by the desecration of a nearby woodlot as by a famine in China, a near-program [pogrom?] in Germany, or the murder of slave girls in Ancient Greece. Individual thinkers since the days of Ezekial and Isaiah have asserted that the despoliation of land is not only inexpedient but wrong. Society, however, has not yet affirmed their belief. I regard the present conservation movement as the embryo of such an affirmation" (Leopold, 1933, 635). J. Rodman suggests the need for a new ethic, asking: "Exactly what is so frightening about the prospective death of Nature?" (Rodman, 1977, 116). R. Routley and V. Routley argue even more explicitly for preservation of species and wilderness at the cost of human interests, human concerns, and even of what their opponents claim to be human rights.

The Routleys and those of like mind would appear to be correct about the intuitive insights of those who live in affluent countries in affluence. Most such people are morally outraged at the wanton destruction of species such as the blue whale, and of wilderness areas such as rain forests, much as moral persons are outraged at serious atrocities involving the deaths of persons even though they appear to be less moved by the destruction of lesser species, rare, ugly butter-flies, ants, plants. Nonetheless, it will be my concern in this work to reject this approach and the moral intuitions on which it rests. Much can be done to explain why it is so obviously wrong to endanger and render extinct certain species such as the blue whale, and to destroy natural phenomena and wilderness areas such as the Great Barrier Reef. However, as the theory developed here implies, ultimately, when there is a clear clash between human welfare and human rights on the one hand and preservation of wilderness and of plant and animal species on the other, human welfare and respect for human rights must prevail, unless what it is proposed to preserve has great intrinsic value, as a beautiful object or a source of knowledge. The case for preservation, and even more, for there being a prima facie presumption in favor of preservation from a concern to conserve actual and possible resources, to promote good, beauty, and knowl-edge, is a very strong one, on the basis of which much endangering of species and destruction of wilderness is to be condemned as morally wrong and morally reprehensible. The reason it cannot be established that all preservation is prima facie morally obligatory and desirable is that not all species possess intrinsic value, since species qua species do not possess intrinsic value, and because only some wilderness, not all wilderness, possesses intrinsic value. Further, independent of reference to intrinsic value, there seems no reason to believe that destruction of species or wilderness is intrinsically wrong or respect for species and wilderness intrinsically obligatory in the way the duties of justice, honesty, and respect for persons are intrinsically obligatory independent of a duty to respect the intrinsi-cally valuable.

Preservation as Dictated by Concern for Conservation of Renewable Resources and Securing the Intrinsically Valuable

A number of distinct, very different arguments for preservation are seen on examination to be arguments for preservation as dictated by concern for conservation of renewable resources for present and future generations of human beings. One argument stresses the irreversibility of destruction of wilderness and species and the unpredictability of the consequences of needlessly interfering with nature in this way; a distinct argument notes that endangering useful species by overkill is plain bad economics; another notes that seemingly useless species, like seemingly useless raw materials, may, in the future, become very important resources or sources of resources; the importance of maintaining wild stock because of the vulnerability to disease of cultivated, "artificially" created varieties is pointed to in yet another argument; other arguments stress the value of wilderness for present and future generations in terms of its psychological and recreational value. These add up to a powerful case for some sort of initial presumption against destruction of species and wilderness unless a strong case is made to the contrary. This presumption becomes strong when these arguments are reinforced by arguments about the intrinsic value of certain species and certain wilderness. We will look now at these arguments and at what they show and do not show.

The Irreversible Nature of Destruction of Species and Wilderness and the Unpredictability, Because of the Complexity and Interconnectedness of Things in the World's Ecosystem, of the Consequences of Such Destruction, Where the Consequences May Be Very Disadvantageous to Man

Clearly, interference by way of destruction of species and wilderness, because it may have far-reaching consequences due to the delicate balances in nature's ecosystem, may be very disadvantageous to man in ways that he in no way anticipates; that it is irreversible may make it even more so, even dangerously so. More, the methods used to exterminate plants and animals, for example by the use of pesticides, or to use or tame wilderness, as by clearing forests or damming rivers, may do great indirect harm. These facts clearly constitute powerful considerations in favor of the utmost caution in interfering with nature in irreversible ways, as by seeking to exterminate species as weeds and pests, or by way of seeking to bring about destruction of wilderness areas. However, they do not constitute a case for never seeking to exterminate a species, for never encroaching on or using wilderness. If the harm done to man by the existence of certain organisms is very great, as it appears to be in respect of disease organisms such as rabies, cholera, the malaria organisms, and parasites such as the tape worm and the hook worm, in the absence of positive evidence to the contrary it may well be right to destroy such organisms. If we can safely preserve living organisms of various of these kinds in laboratories and exterminate them in the world at large, that may well be preferable. However, if they cannot be preserved with safety in laboratories, or if they do not survive in such environments, it may be right and obligatory for man to eradicate them if he can.

This conclusion is contested by many preservationists. Consider B. Commoner's argument in terms of his "third law of nature," set out in a weaker form as "Any major man-made change in a natural system is likely to be *detrimental to* that system," and in a stronger form as "Nature knows best"; note also his first law, "Everything is connected with everything else" (Commoner, 1972, 37, 41, and 29). These contentions are supported by reference to his general claim about stability, balances, and equilibriums in nature's ecosystems, and by an analogy between nature and a watch, along the lines of the well-known argument for God's existence from the evidence of design in the universe, the analogy running: "In other words, like a watch, a living organism that is forced to sustain a random change in its organization is almost certain to be damaged rather than improved" (ibid., 38).

Such an argument is too good to be true. It implies that we should leave it all to nature. It ignores those features of nature that have led to descriptions of nature in such terms as *harsh, savage, cruel, ruthless.* Given man's seemingly successful use of his environment, first in hunting, then in agriculture, then in developing new varieties of plants and animals, then in developing industries, cities, civilizations, culture, science including medical science, technology with which to combat the "ravages" of nature, Commoner's claim appears to be blatantly false. These developments all represent massive interferences with nature. If even one such development rendered the earth a better place for man it would falsify Commoner's claim. In fact, most appear to have made a better life possible for man. Commoner's claim then cannot mean that nature knows best for man. Yet it is unclear for whom or what it is being said to know best. Perhaps, as is suggested by the weaker version of the "law," it is being claimed to know what is best for itself. If this is what Commoner's claim amounts to, the reply is obvious. Why should man accept what is best for nature?

The empirical facts falsify Commoner's claim when it is judged in terms of any acceptable standard of values. Nonetheless it is worth considering how a scientist might have come to offer such a view as a finding of science. The thinking behind it, with the use of the analogy of the watch, calls to mind Paley's well-known statement of the argument from design, an argument that is generally discredited in the eyes of scientists who believe that the facts and the seeming evidence of a wise designer and excellent design are better explained in terms of evolutionary selection. Further, as Dubos has observed, there are other grounds for empirically rejecting Commoner's claim, namely, that even in terms of Commoner's seeming criteria of "best," man can improve upon nature by enriching its diversity.

Commoner's law, and the attitude toward interference with nature that it exhibits, is neither scientific nor rational. It amounts to a less than frank appeal to irrationality, to the kind of conservative irrationalism about planning on the basis of a rational, informed assessment of the likely consequences of one's action and inaction that underlies Edmund Burke's conservativism, the more recent conservatism of M. Oakeshott, and the laissez-faire liberalism of Herbert Spencer and others. To abandon planning, albeit fallible yet carefully thought-out planning on the basis of our knowledge and our awareness of its limitations, to allow cholera and bubonic plague, rabies and malaria, to run unchecked because "nature knows best," because we cannot be certain that due to the interconnectedness of things in nature we may do more harm than good by engaging in quarantining humans, animals, plants, in killing off mosquitoes, fleas, rats, wild dogs, and in using vaccines and appropriate drugs and antibiotics to combat diseases and antivenin to check the effects of animal poisons, to use only wild stock of plants and animals, and not to domesticate the latter, is to abandon morality and the moral point of view for irrationality and the abdication of moral autonomy. Having so argued, it is to be noted that the case from disease is not quite as strong as it may at first seem to be. Although preventive medicine, surgery, and the like have made obvious advances in the fight against disease and there appear to have been no major adverse effects resulting from the use of vaccines—including such vaccines as that against smallpox which is no longer necessary because that disease has apparently been conquered—by the use of antibiotics we are creating more virulent, resistant strains and could find ourselves extremely vulnerable to such new strains if new antibiotics are not found to combat them as they arise.

Acceptance of Commoner's attitude and Commoner's law would also involve gross acts of immorality, the immorality of withholding medical aid from the sick and vulnerable, from those with diabetes and other genetically based diseases. It could also lead to ecological disasters, allowing nature to take its course in respect of human population growth, no matter what the size and rate of growth of the world's population. The use of artificial methods of contraception constitutes a violation of nature, in terms of Commoner's position an impermissible one, no matter what the state or size of the world's population.

The other side of the picture regarding the dangers of interfering with nature, endangering species, and encroaching on wilderness, lies in the fact that Dubos notes—namely, that nature is very resilient and it is often much harder to damage wilderness or eradicate a species than is suggested by Commoner's claim. Dubos points to the rapidity with which flora and fauna recovered from the effects of the Krakatoa volcanic eruption. The same phenomenon is evident with the less devastating volcanoes of this century. Consider also the considerable adaptability of most forms of life, how mosquitoes become resistant to pesticides, rabbits to myxomatosis, how algae grow luxuriantly in contaminated waters, and how even human beings multiply and survive in conditions that seemingly on objective criteria are too poisonously polluted to permit human existence. That it is so difficult to eradicate unwanted organisms such as rats, black-berries, bracken, and the like testifies to nature's adaptability and resilience.

The conclusion to be drawn then is that although the limitations of our knowledge, the irreversibility of eradication of species and de-struction of wilderness, and the fallibility of our predictions about the effects of our interferences with nature constitute grounds for care-fully assessing both the desirability and necessity of acting not to preserve, they do not constitute overriding grounds for always opting for preservation. It can well be our duty not to preserve certain species and to encroach on and use for recreation or agriculture areas of wilderness. That it is possible that there already is no real wilder-ness left in the world, even early man having a great impact on his environment by his use of fire, further supports this general qualifica-tion.

A second important argument is that to exterminate a species is often bad economics, bad management of the resources of the world, rendering renewable resources nonrenewable and ultimately nonex-istent. Consider whaling, fishing of various fish, some hunting as of certain varieties of kangaroo. They illustrate G. Hardin's tragedy of the commons, the self-defeating nature of the unchecked, short-sighted pursuit of individual self-interest. Fortunately they would seem not to be the norm. If we adopt what must at present stage of our knowledge be a questionable view, that whales cannot be posses-sors of moral rights and hence may legitimately be used as resources by man for food and other products, then the conduct of the whaling industry in this century, more evidently in respect of overkilling the blue whale, has involved irrevocably destroying what could have been a continuing, renewable resource, and this for the sake of no long-term gain. In terms of resources, the world is worse off for the extinct and endangered whale species. And man could in other ways be worse off than he need be for the loss of these species. This is true also of the rendering extinct of many other species of animals and plants that could have been used and yet conserved but which, for the sake of small short-term gain by a few individuals or nations, have been overexploited and permanently lost to man. Similar wasteful destruction of natural resources has also occurred in respect of wilderness, usually thereby also endangering rare plant and animal species. Wilderness, unsuitable for agriculture, has been converted to agricultural use only to become desert. Rain forests have been cleared for timber only to become wasteland. Man has thereby lost what was, and could have remained, a resource.

It is also true and important that many species that appear not simply to be useless but to be positively harmful to man may ultimately be found to be very valuable, for example as sources of foods, medicines, counteragents for other weeds and pests, or for producing commodities of various kinds, and that therefore we ought not to seek to eradicate, endanger, render extinct such species, given the irreversible nature of such an action, unless there are compelling reasons for doing so. Many so-called weeds and pests have proved to be significant resources. Uses are continually being found for what seemed useless and harmful. Even such unwanted weeds as the water hyacinth and Paterson's Curse are being put to use (for paper and honey respectively), although it remains to be seen whether the uses to which they are put can outweigh the harm they cause.

The basic claim upon which this argument rests could be of sufficient truth and relevance for it to be prudent for mankind to seek to retain in escape-proof laboratories, herbariums, and zoos, such dangerous, harmful organisms as smallpox, rabies, malaria, and the like, as well as useless, dangerous plant and animal species, if we ever succeed in eradicating them from their natural habitats. Protection of human beings from harm or nuisance from such species rarely calls for total eradication. However, it may call for action that may endanger the existence of the species. Many animals do not flourish in zoos, plants in herbariums, or organisms in laboratories.

Important economic and biological arguments are advanced in terms of the vulnerability of cultivated species of plants and animals to disease and injury, and of the importance of maintaining wild stock, and, more generally, the broadest, most complete gene pool we can. The many improved varieties of wheat, rice, maize, and potato that have been developed during the past 500 years, including the so-called wonder strains of rice and wheat of the Green Revolution of recent years, which have led to the production of vastly more wheat, rice, maize, and potatoes, and to production in areas that otherwise would have been unproductive and under conditions that formerly would not have allowed production, are vulnerable to disease in a way the native stock is not, even though it is true that artificial varieties are bred to be resistant to specific diseases. Geneticists are aware of the possibility of serious problems due to diseases striking new strains, as the potato blight did the potato crops of Ireland in the mid-nineteenth century. Not to preserve the wild species is to court needless dangers. The same is true of improved

varieties of oxen, sheep, pigs, goats, horses, dogs, cats. This constitutes a major prima facie consideration for preservation. However,
without further argument, such a contention would not provide
grounds for preserving different, dangerous species, as many as
possible, species such as those of venomous snakes which do significant, determinate harm and no evident good that cannot be achieved
in less costly ways. Indeed, plausible reasons are needed, given the
very evident benefits to be achieved by eradicating them from their
natural habitats, for not doing so. Nonetheless, given the argument
from their possible value as a resource—medical uses are being found
for snake venom—it would seem to be prudent for mankind to ensure
that such animals survive in zoos if, as seems to be the case, it is
possible to preserve these species in this way.

Arguments for preserving wilderness have been developed in
terms of its value as a resource that provides for man's psychological
needs and recreational interests. People are said to gain psychological
uplift from wilderness; it is said to fill a void created by life in a
technological society; and persons derive pleasure from visiting wilderness areas, more evidently those that are beautiful. The psychological argument, while seemingly true in respect of some persons—
there must be a doubt since we cannot be sure that what appears to us
as wilderness really is such, given the interconnectedness of nature
and man's impact on nature from the earliest times—is evidently false
as a generalization about all human beings. Many, probably most,
Europeans manage very well without encountering wilderness; they
find non-wilderness areas such as the European Alps, the Scottish
Highlands, the English Lake District, the Rhine valley, and the like,
give them the satisfaction that others claim can be obtained only from
wilderness. Most city dwellers throughout the world probably never
encounter wilderness, although many get satisfaction from visiting
country areas. Many are positively frightened of remoteness, and of
and by wilderness. The argument that wilderness is a recreational
resource must be a very qualified one, one that can relate only to the
recreation of a minuscule percentage of mankind. The more wilderness is used for recreational purposes, the less wilderness it is, and
the more endangered it becomes as such. Visitors disturb fauna and
flora and may unknowingly, unintentionally, even with the most
stringent quarantine checks compatible with its use for recreation,
significantly change the equilibrium of its ecosystem by introducing
new organisms or changing its physical character. Such a defense of
wilderness is elitist in a matter in which the case for elitism is weak.
Of necessity, only relatively few (and since it is impossible to judge
need fairly here and to discriminate on the basis of it), the most needy
will not necessarily benefit. Yet the use of wilderness areas for
agriculture may benefit millions of needy persons.

Other arguments relate the preservation of nature with human
well-being in less direct, simple ways. Dubos argues that man, being

himself a product and part of nature, must maintain contact with his origins: "Technology is giving man immense power over the cosmos, but in its present form it is depriving him of the sustenance he could derive from direct contact with nature. . . . Because man is still of the earth, he too loses attributes essential to his survival when he allows the technological way of life to dissociate him completely from the natural environment" (Dubos, 1973 13–14). And: "Man is still of the earth, earthy. The earth is literally our mother" (Ibid., 38). And: "The immediate danger is not the destruction of life but its progressive degradation. . . . But they fear that if present trends were to continue the physical environment would become progressively impoverished in sensual qualities and the social environment would have to be so highly organized and regimented as to resemble that of social insects" (ibid., 219–222). J. Rodman, in like vein, writes that "human nature as grounded in an order of biological diversity" such that "every extinction of a species seems an impoverishment of human life"; he observes earlier in his discussion "that a totally humanized world would diminish us as human beings," and asks, "Does it not indirectly affirm the intrinsic value of the nonhuman realm, or at least suggest that their loss in either realm is a loss of value?" (Rodman, 1977, 117, 116). As against Rodman's own suggestion that the considerations to which he points affirm the intrinsic value of the nonhuman in nature, I suggest that they support a qualified version of the second alternative, and hence, that concern for nature for these reasons is overall in man's interest because of his nature as an organism and a person. These discussions, so understood, therefore suggest that species, wilderness, nature, are to be valued for human-centered reasons, because they fulfill a need, open possibilities, because in other ways they are vital to human self-expression, human self-awareness, human dignity, and the emergence of those capacities and creative endeavors that man rightly values.

In brief, in terms of these and other arguments that argue the case for preservation from a consideration of conservation of actual or potential renewable resources, a strong case for preservation can be developed. However, there obviously will be situations in which concern for human well-being and human rights will dictate that species be endangered, wilderness encroached upon.

The duty to promote good provides a basis for a strong case for *selective* preservation of wilderness and species. (Equally, the duty to lessen evil might be used to justify selective interference with wilderness, selective destruction of species.) The duty to promote or secure good would dictate that beautiful wilderness areas such as the Great Barrier Reef, beautiful animals such as the tiger, the blue whale, the lyre bird—the individual creatures and not simply the species—be protected and even multiplied in numbers, unless some more stringent obligation overrides this duty.

Further, given that knowledge is rightly commonly also seen to be an intrinsic good, there is a significant case for preserving all species in their natural habitats, to be developed from the intrinsic value of full and continuing knowledge. This implies that wilderness as the natural habitat of many species is also to be preserved as a condition of such knowledge, as well as as an object itself of knowledge. Again, this is simply a prima facie duty, which may be overridden by more stringent duties such as the duties to protect human life or alleviate suffering.

It might be thought that the duty to promote good could be extended to justify the preservation of all plant and animal species in terms of the intrinsic value of all life, or of all animal life, or of all species and subspecies. Against this, I suggest that such a mode of argument cannot succeed, that it is significant that no ideal utilitarian has attempted to extend ideal utilitarianism along these lines. There is no evident value in life as such. Those who, like Albert Schweitzer, have sought to argue to the contrary, have nothing convincing to offer against this view. Only certain kinds of life have value, and then the value derives from the possession of valuable capacities, rationality, including imaginative, emotional, intellectual capacities. (See McCloskey, 1979, for a fuller discussion of this issue.)

It is often suggested, but much less often explicitly claimed and defended, that species qua species possess intrinsic value. R. Lamb is one who ventures to make this claim in a critique of my paper, "Ecological Ethics and Its Justification: A Critical Appraisal" (Mannison, McRobbie, and Routley, 1980, chaps. 6 and 7). Such a contention is implausible once it is appreciated how species are characterized, and once it is realized that those who argue for preservation of species usually mean their arguments to apply not simply to species in the strict sense but also to subspecies, kinds, subkinds and so forth. A species is characterized in terms of uniqueness in respect of the capacity to produce offspring that can themselves reproduce; hence, species are explained in terms of uniqueness as a gene pool. Those who argue for preservation of species usually mean species in the looser sense of a unique kind or variety. Thus in both senses of species, the key feature is the relational one of uniqueness. Like the related attribute of rareness, uniqueness is relational and depends on what else exists, and is not an intrinsic property connected with other constituent properties.

There are real difficulties in the way of claiming that intrinsic value attaches to a species qua its uniqueness and not its various attributes and properties, be it the blue whale, the black snake, the tapeworm, the malaria organisms, the mountain ash, the species of clover, or fern. It is easy to see why some of these species are valued. Some, such as the black snake and the mountain ash, possess a rare beauty. Some, including the blue whale, the mountain ash, and the species of clover, have commercial, economic, or utilitarian value. However, in claiming that there is intrinsic value in the continued existence of these species, something other than the aesthetic or economic value is being claimed to exist. To assess this claim, it is easier to take examples of species that lack both aesthetic and economic value, which are ugly and uneconomic to man. Consider the tapeworm and the house fly. The tapeworm is ugly and repulsive although beautifully adapted to its role; the fly is neither beautiful nor useful. Why then seek to maintain such species in existence? There seems to be nothing worthwhile about them. They could be valued, if at all, simply for their uniqueness and not as species that are valuable because they possess certain attributes, colors, shapes, features, organs, capacities, and the like. It is simply because the species is unique in a very curious respect, that of producing fertile offspring, not because it possesses certain specifiable characteristics, that it is valued by those who value species. This is confirmed by the fact that such ecological ethicists who value species qua species are committed to valuing any species whatsoever, whatever its characteristics. This means that it cannot be the characteristics which are the basis of the intrinsic value attributed to species, but the uniqueness of the species. Yet to value uniqueness is to do something very strange. It is rather like valuing rarity. With properties such as rarity, the property ceases to hold of a thing, if many identical specimens are found; the existence of the other specimens would take away the rareness and hence the value that the rare thing possessed. With species and uniqueness of species, this of course cannot occur, since it is logically impossible to find an identical species. The species that is identical in the relevant respect of reproduction (and/or whatever else may come to be the defining feature of species) but which differs in many other attributes would not be regarded as a new species but as a variety within the same species. Hence, what is being valued in valuing species is the species-characterizing uniqueness, reproductive uniqueness. Other attributes may differ greatly between members of the species, but the species possesses value qua species. A curious implication of such a valuing of species is that it is not the members of the species as individuals that are valued, but members as embodying the species. Respect for the intrinsic value of a species demands simply that it not be endangered, not that its numbers not be decimated.

In presenting these arguments for preservation from man's good and from the duty to promote good, no mention has been made of the fact that nature does not have a fixed number of species that will continue in existence unless man destroys them. Species have been coming into being and going out of existence ever since life began on earth. The rate of coming and going appears to have been greatest before the evolution of man. It is therefore unclear what ultimate effect results from man rendering a species extinct, in particular, whether man permanently reduces the diversity of species or simply leaves room for new species to come into being, or whether his destruction of species is only of minor relevance to the total number of species the world will contain in the future.

Further, if mere number of species were what was of importance, and this would be so if species qua species possessed intrinsic value, man could artificially seek to multiply the number of species, as by changing the environments of various species; or he could work to create new species by genetic engineering. In fact, there would seem to be no gain in intrinsic value in the universe if man succeeded in creating new species in these ways, unless the species were possessed of beauty or some other valuable attribute. The value of genetic engineering, if it does enable man to create new species, will not lie in the mere creation of new species—that could be valueless. It will lie in the nature and value of the actual species created. For genetic engineering to create useless, harmful species, species of flies, fleas, mosquitoes, lice, would be for it to do something that in no way enriched the earth.

The foregoing are important arguments relating to preservation. They establish not that we must preserve species and wilderness at any cost, but that it is often morally wrong and very imprudent not to preserve species and wilderness, and that we need to have good moral reasons for not doing so. Many ecological moralists are dissatisfied with such a qualified defense of preservation and seek to make out a stronger, less qualified, less human-centered case. The following are among some of the more important arguments of this kind that are to be encountered either implicitly or explicitly in many ecological moral writings. Given their evident unsoundness and inadequacy, that they are advanced as being important arguments tends to have the effect of undermining and weakening the impact of the foregoing arguments.

Thus it is often suggested, although less evidently explicitly argued, that *a world with a diversity of species is better than one with few species*. Such an argument attributes value to diversity as such, or to diversity as a condition for the occurrence of other goods. This is far from self-evidently true. A world with double the number of species the world now contains would not evidently be a better, more valuable world than that we now know. If we take evolutionary theory seriously as an account of how the species that now exist came to exist and to survive to the present day, and why they will cease to exist when that occurs, we should have no ground for supposing that it was because they possessed value as contributing to the value of the whole of which they are components. Rather, we should conclude that they exist and have survived because of contingent facts about their survival capacities in their present environments. With changes in the environment, some will lose their capacities to survive, others will evolve such that those with survival capacities will survive. Over the millions of years during which life has existed on earth, species have been coming into existence and going out of existence. Man may recently have accelerated the rate at which species go out of existence, but natural climatic and other changes could have vastly greater effects than has man even in recent years. The coming of an ice age could be more destructive of species than are pollution, pesticides, or agricultural development of wilderness areas. Whatever is true in respect of the present day, it is evident that over the whole period of his existence man has been only one relatively minor factor bearing on species survival rates. Given that the capacity of species to survive is continually being tested, man is simply adding more hazards to those created by nature itself for some species. We tend also to notice the species that are most obviously adversely affected by man and not to notice those whose continued existence is assisted by man's effects on his environment. Yet there are many such species.

In the looser sense of *species* meaning variety, subkind of some sort, man has greatly enriched the earth by breeding useful and beautiful new varieties of animals and plants—cattle, sheep, horses, goats, pigs, dogs, cats, turkeys, fowl, rice, wheat, corn, potato, roses (and many other flower varieties), as well as many varieties of fruits. With the advent of genetic engineering, it can confidently be anticipated that man will create or cause to come into being many new varieties or species; what is of importance is that they be ones which improve the world, ones that are useful and beautiful, not harmful or dangerous.

At this point the preservationist may seek to argue that it is not sheer number and variety of species and subspecies that enrich the universe but the special variety that nature has provided, a changing, evolving variety, that has special value. Given the evolutionary account in terms of natural selection of species, this seems highly unlikely. This kind of claim can reasonably be supported only by moving away from a scientific account of the origin and selection of species to a more religious pantheistic or mystical view of nature.

The pantheist view of nature, of nature as a living whole of which we are parts or aspects, or of mind in action, is also suggested by the popular way of talking about the natural world in an anthropomorphic way. It is unnecessary to rehearse here the arguments that have been pressed against pantheism in its manifold forms. It is sufficient to note the problems in the way of the claim that the universe is a unity. All we have is evidence of interaction, of temporary, changing links of interdependence between different elements in the universe. There appear to be no good reasons for deifying nature, personifying it, seeing it as having a will that we should respect and with which we should align ourselves.

Preservation of Nature and Noninterference or Minimal Interference with Nature Has Also Been Urged on the Basis of Quasi-scientific Laws, Quasi-religious, Religious, and Metaphysical Beliefs about Nature

In addition to the theories to be discussed here, there are or could be theories that are purely theological, theories of revealed religion, and the like, to the effect that God has so arranged the universe and man's place in it, in respect of other living beings and inanimate nature, that he has imposed on man duties to care for some or all of the things in nature, for their, not man's, good. Discussion of such theories would take us into the whole question of the existence of God, our knowledge of his mind and will. If this kind of theory could be filled out and shown to be true, it would be of great relevance to ecological ethics and could provide the basis for a God-centered, not man-centered, approach to nature and to respect for nature. However, since no significant religious position and arguments supporting such a position have been developed—the orthodox theories are that nature is for man's good and man's use, he having duties of stewardship, if at all, to care for nature for man's sake—and since, for reasons I have developed elsewhere, I reject the belief that God exists, it is sufficient here to note that this could be an important line of defense of the view that nature should be respected, preserved, cared for, even at the cost of human well-being, if the relevant arguments could be developed.

Among the theories to be examined here are those that proceed to ethical, practical, moral conclusions from a belief in: (1)*the existence of a natural harmony in nature, a natural balance or equilibrium* which man, by his intervention, may destroy (biotic pyramids figure as part of this account with Leopold and others); and (2)*the interdependence of things,* which is such that whatever interference results from man in nature will have *unpredictable, harmful effects, nature acting wisely and beneficially, if not thwarted by man.* Theories developed here are either scientific theories—they are commonly advanced as if they are such and as if they possessed all the authority of well-grounded scientific laws, in which case they would be testable in principle—or they are advanced as metaphysical or religious theories, theories that do not need and do not admit of empirical verification and falsification. Theories of the second kind move more evidently into becoming versions of pantheism. Various ecological moralists move between the range of positions noted here without realizing the range and kinds of positions they move between are so various as they are.

Clearly, if advanced as scientific laws, generalizations, theories, the above need to be explained as explanatory hypotheses, reports of regularities, causal laws, or the like. Hence we should need to know what such theories explain and explain better than any alternative theories, what their predictive value is, what counts as confirming evidence, and what, if anything, would count as disconfirming evidence, and also, whether man's actions in shaping nature, like the "actions" of the buffalo, beaver, swan, rabbit, ape, are natural and in accord with the law of nature, whatever be the character of man's actions, or whether they are to be seen as outside nature, wholly or partly so, and if so, why. If man is deemed to be an insider, part of nature, nothing he can do can be unwise, wrong, contrary to nature. If he is deemed to be outside nature, it would need to be shown why this is claimed to be so, and also that his intervention is as claimed, always, sometimes, wrong, unwise, contrary to nature, when and why it is such, and this in terms of scientifically acceptable criteria of "harmful," "wrong," "unwise." Thus when man breeds a new, beautiful variety of rose, horse, or sheep, a more productive variety of rice, wheat, or potato, when he prevents a species from being eradicated because of a natural disaster, a volcano, earthquake, tidal wave, or pest plague, he is always acting wrongly, unwisely, harmfully, because he is acting contrary to nature. Is it that it is thought he is acting in opposition to evolution when he so acts, and that he ought to align himself with evolution even when this means not saving an endangered species from destruction by natural forces? If so, what possible reasons could there be for such a view? Other problems in respect of such claims relate to the lack of scientific precision in the formulation of these laws and theories. They are commonly so formulated as not to admit of scientific scrutiny and assessment. Further, where the laws incorporate a value term used evaluatively, terms such as *best*, the difficulties in the way of establishing that the laws are scientific ones are insuperable. Yet, if such terms as *best* are used descriptively the laws in which they appear fail of their purpose.

The contention that there is a natural harmony in nature, that man, by his actions, endangers this natural harmony, and that it is a bad thing for man to act to endanger or destroy this natural harmony, may now be considered. The picture portrayed here is that the universe—or the earth and its environs—is one immense ecosystem composed of a hierarchy of ecosystems, such that their components come to be absorbed into other ecosystems. Men can disturb the equilibriums of ecosystems—they are in a continual state of adjustment and readjustment due to factors other than human actions—so that new equilibriums develop, or they can so disrupt a lesser ecosystem as to destroy it and cause a reformation of ecosystems. Few suggest that man can or will destroy the ecosystem of the earth or of the universe. Rather, it is suggested that he will bring about a change in it, such that it is a lesser ecosystem than it would naturally be. What is important here is that nature is not a set of fixed harmonies, balances, equilibriums, but a fluid series of processes of adjustments. Man causes different adjustments, different systems to develop.

This means that the heart of this kind of claim is that man destroys valuable, in a nonhuman sense, ecosystems by his actions, and that the resulting harmonies, equilibriums, balances, are also less valuable. It is man's intent when he converts wilderness to agricultural use to alter the ecosystem, to change the nature of the equilibrium that will prevail. He does so because he thinks it is better to have that sort of ecosystem. In what way is he mistaken in his belief simply because his ecosystem and its equilibrium are distinct from those of nature?

Man clearly does change the ecosystems within the world. He seeks to do so when he builds dams, engages in irrigation, straightens a river, takes action to prevent flooding, reclaims waste, marsh land as in the Wash, kills disease-carrying pests such as fleas and some of their hosts, rats, mice, and the like. Of course, man destroys the ecosystem that exists between himself as host and the debilitating parasites that feed on him, if he kills the parasites. What needs to be shown is that the new ecosystems, with their new equilibriums, are less valuable than those they replace. Most of us would, without hesitation, say that an ecosystem that contained a deloused army was a better one that one that contained a lousy army, that an ecosystem from which rats and their plague-carrying fleas had been removed was a better ecosystem, although no doubt, if they could have views about the matter, the lice and rats would dissent from this view.

Man, of course, makes mistakes in his judgments. In retrospect, it would seem that man could have used DDT to better effect. The Aswan Dam had not proved the unequivocal boon it was expected to be. Equally, ecologically speaking, it is early days even now to assess its ecological impact, gains and losses. Against such cases, there are many unequivocally beneficial changes in ecosystems, replacement of ecosystems, changes in their equilibriums, that have come about through the construction of dams, irrigation systems, the use of pesticides, fertilizers, and equally, through family planning using many methods of birth control. Thus, if the general claim is rendered a factual one, one explained in terms of agreed criteria of good and goods, it is evidently false in regard to numerous interferences. All the scientific claim can come to is that man causes changes to and within ecosystems. A value judgment of a very difficult and complex kind needs to be made to determine whether these changes have been beneficial. With the wisdom of hindsight, we can see that they could often have been more beneficial and involved less destruction of species and wilderness, less irreversible damage than they did. Grave mistakes have been made and no doubt will be made in the future. However, it would be a bold person who claimed that overall man has failed on balance to modify the earth's ecosystem to man's advantage as opposed to his best advantage, that man would really have been better off if he had allowed nature outside himself to take its course.

If it is conceded that the latter was never a serious moral option for man, the question becomes not, Should man seek to avoid changing nature's ecosystems? but, How, in what ways, should man change nature's systems? To write as if nature always works for the best, as if all existing ecosystems and their equilibriums are as they ought to be, is to distract attention and thought from the real problems that confront mankind. Man's primary problems must relate to determining what are the desirable interventions, the beneficial changes to be made to nature's ecosystems.

The difficulties that confront the kinds of theories we have been considering, if they are initially advanced as factual, scientific theories, as products of scientific, even of, ecological wisdom, scientific research, scientific expertise, lead to them being advanced instead, implicitly if not explicitly, as metaphysical or religious truths. Commoner, by his use of the analogy of the watch, betrays a religious as distinct from a scientific attitude toward nature, one of a kind that has led to explaining nature in terms of it being a product of a supreme designer, not of evolutionary selection. It is not my purpose to claim that Commoner embraces a theistic account in *The Closing Circle*, simply that his position would be more intelligible and coherent if he had done so. Clearly, pantheistic and theistic, teleological theories that there is an innate purpose in nature due to some inner mind or cause that confers on the operations of nature a genuine purpose would underpin claims such as Commoner's; they would of course need philosophical and ethical backing. In the absence of such arguments, and given that the theory of evolution adequately accounts for what formerly was explained in terms of the work of a wise designer, there seems no reason to give credence to such theories.

Holistic, organic theories of nature and of the universe, where nature, and/or the universe is/are seen as organism/s with connected, interdependent parts sharing a common life are not uncommon in ecological ethical and political writings, although they more usually are implied rather than explicitly stated and developed. C. D. Stone's qualified espousal of such a view in terms of a myth, but a special kind of myth, is not untypical. Stone observes: "What is needed is a myth that can fit our growing body of knowledge of geophysics, biology and the cosmos. In this vein, I do not think it too remote that we may come to regard the Earth, as some have suggested, as one organism, of which Mankind is a functional part—the mind, perhaps: different from the rest of nature, but different as man's brain is from his lungs" (Stone, 1972, 51–52). Those who are attracted to this view obviously see the same kinds of ethical conclusions to follow from the relationship of organs to their organisms as did Hegel and the Hegelians, even though Hegel's organisms were of a very different character.

The attractiveness of the organic theory lies in its ready accommodation of the interconnectedness of things, and hence in explaining why seemingly superficial interference with what appears to be loose and separate may have major and damaging repercussions. It is also thought to establish the moral priority of nature as an organism vis-à-vis its organs, especially man.

This mode of seeking to justify veneration of nature at the expense of man's well-being cannot succeed in its object; it does not provide a justification of preservation at the expense of human welfare and the enjoyment of human rights. This is because it is not self-evidently the case that nature is an organism, that every element in nature is interdependent, and interdependent in the relevant ways with every other, except in the very trivial senses of having spatial, temporal, and varying casual relationships. There is no adequate evidence available to support the view that the world is a unified organization of all things, animate and inanimate, around a common life. There is no common life. There are problems in the way of elucidating the concept of life in or of nature as a whole, as distinct from the life in various of its members. These issues need not be explored further, as clearly the onus of proof is on the organicist to show that nature is an organism with a life of its own, and that this life ought to be respected.

The available evidence shows that at most all that can plausibly be argued for is that nature is like an organism. It is and it is not. Where there is only likeness and not full identity, there must be differences. The other element of the organicist's claim is also mistaken. Nothing of any importance morally follows from the mere fact that something is an organism, and something else simply an organ of the organism. Organicists treat it as self-evident that the organism is in some way higher, better, superior to its parts, the organs being in a naturally subordinate relationship with the organism. If we consider actual organisms, plants and animals, we find that the ethical conclusions of the ecological ethicists do not hold there, although it is true that there is a biological dependence between the organ and the organism, with most organisms and most organs. Clearly, the organicist must argue from the precise nature of the organism, nature, and from the demonstrable harmfulness of the interference with the organism, and also from the value of the organism itself, if he is even to hope to justify the condemnation of interference with nature. This is precisely what he is unable to do. That is why he invokes the analogy of an organism.

A New, Specifically Ecological Ethic

The conclusions to be drawn from the preceding discussion are that much destruction of wilderness and much endangering and eradication of species of plants and animals are imprudent and morally wrong, as destroying both what is or may be of value to man, to those now living and to those of future generations, and/or of what is intrinsically valuable. It is immoral as showing lack of respect for persons, as failing to promote good and lessen evil, and it is often also unjust, as apportioning to those living today a disproportionate share of the earth's natural resources in living organisms and wilderness. It is imprudent as unnecessarily irreversibly destroying renewable resources that may be needed in the future. The foregoing arguments make out no case for preservation for its own sake. The attempts to show that, which have been examined up to this point, have been seen to be inadequately and unsoundly based. If these are all the relevant moral considerations, it would follow that when human interests and concern for human well-being and human rights unequivocally dictate the destruction of wilderness or the endangering of plant and animal species, we have both the right and the duty to act out of concern for well-being and human rights.

It is awareness of this that has led to the now common claim that we need a transvaluation of values, new values, a new ethic, an ethic that is essentially and not simply contingently new and ecological. Closer inspection usually reveals that the writer who states this does not really mean to advance such a radical thesis, that all he is arguing for is the application of old, recognized, ethical values of the kind noted here under the characterization of respect for persons, justice, honesty, promotion of good, where pleasure and happiness are seen as goods. Thus, although W. T. Blackstone writes: "We do *not* need the kind of transvaluation that Nietzsche wanted, but we do need that for which ecologists are calling, that is, basic changes in man's attitude towards nature and man's place in nature, toward population growth, toward the use of technology, and toward the production and distribution of goods and services. We need to develop what I call the ecological attitude. The transvaluation of values which is needed will require fundamental changes in the social, legal, political and economic institutions which embody our values" (Blackstone, 1974, 17; see also Blackstone, 1979), he concludes his article by explicitly noting that he is not really demanding a new ethic, nor a transvaluation of values. E. P. Odum's "Environmental Ethics and the Attitude Revolution" in the same collection expresses a less clearly stated attraction toward but ultimate non-acceptance of the demand for a new, an ecological, ethic. There are, however, genuine demands for, although rather fewer genuine attempts to formulate, a new, an ecological, ethic; the more important such appear to be those of A. Leopold and R. Routley and V. Routley. Before examining their approaches to the development of a new, ecological ethic, it is worthwhile first to consider why there is such a strong and widespread feeling among ecological moralists and environmentalists that such an ethic is needed: what are the facts, beliefs, attitudes, judgments, and concerns it is expected to explain, illumine, and justify, and how it is expected to do so.

A new, ecological ethic is sought to explain and provide a justification for common moral intuitions along the lines that the wanton destruction of the natural is evil, that nature should be respected, that life in all its forms should be valued, that natural phenomena are proper objects of wonder and awe, that the natural has value independent of man and human valuation and appreciation. It is seen as being necessary to justify and underpin the belief that nature does not exist simply for man's use and enjoyment, that it has value in and for itself, that its value predates man's existence and will continue long after *Homo sapiens* becomes extinct. It is an ethic that is expected to explain the moral revulsion so many experience at the wanton destruction of species, as when butterfly collectors after collecting their specimens destroy all the remaining members of a species, when "sportsmen" hunt and kill rare species for the mere pleasure of the sport. It will also show that natural phenomena, the Great Barrier Reef, Ayers Rock, the rain forest, and the like should be respected and not vandalized or damaged or destroyed, in the way that notable human artifacts, be they the Parthenon, Stonehenge, or the *Mona Lisa*, should also be respected. Thus a new ecological ethic must explain valuation independent of human beings, human valuing, human attitudes, preferences, needs, wants; and it must explain what is the appropriate moral attitude toward nature, and why it is such. Such an ethic is seen to be needed because traditional ethical theories fail to do this. Many are human-centered both in their accounts of the nature of valuation and in their accounts of the value of what is to be valued. Those that are not such still fail to explain the respect seemingly due to nature and the natural.

Aldo Leopold

Aldo Leopold was a preservationist who did make an explicit and serious commitment to a new, ecological ethic. In both his "Conservation Ethic" and his "Land Ethic," he unequivocally stressed the need for a new ethic. There he argued, but failed to explain, that we should enlarge the ethical community to include the land and all its inhabitants and contents in the moral community; he saw this as akin to the extension of the scope of morality that occurred when slaves were seen to be and accepted as full members of the moral community. Hence his claim: "All ethics so far evolved rest upon a single premise: that the individual is a member of a community of interdependent parts. . . . The land ethic simply enlarges the boundaries of the community to include soils, water, plants, and animals, or collectively, the land." This contention is developed by reference to the further claim that "a thing is right when it tends to preserve the integrity, stability, and beauty of the biotic community. It is wrong when it tends otherwise" (Leopold, 1966, 219 and 240).

These are very puzzling observations, yet Leopold seemed to believe that it was unnecessary to explain what he meant and why he believed his claims to be true. They are all the more puzzling when it is realized that they must be interpreted in the light of his career, and chair in game management, and his stress on the economics of conservation. (See for example Leopold, 1934.)

One possible interpretation is that we should treat all human beings and all things in nature as part of the ethical community as beings and things that are to count morally. This sounds meaningful, reasonable, impressive, and morally generous. It broadens the ethical community by increasing its membership from human beings alone by adding not simply all sentient beings as an increasing number of moralists now do, but by including the whole of nature. Such a proposal appears less meaningful, less reasonable, less generous, and less impressive when one tries to grasp what it might be for an amoeba, a thistle, an ant, a rock, a river, a mountain to count and to be members of the ethical community as fully as persons. When utilitarians suggested that animals should count, and count equally with human beings, they went on to explain that this was so only in respect of pleasure and pain. This made their meaning clear. Of course, it did not prevent their theory leading to morally abhorrent conclusions, that morally it was indifferent when equal amounts of pain were to be eliminated, that they be those of a dog or a child, or whether it be achieved by the death of a cat or a person. As the utilitarian explanation brings out, things count and are counted on account of certain features.

Perhaps, then, what Leopold meant was that all things in nature should count equally in respect of those aspects that are ethically important. The difficulty with that interpretation is that at present we see to be ethically important such things as respecting a person's life, his will, his well-being, and his rights. If we extend the membership of the ethical community, we should, on this interpretation, extend to the new members respect of the kind extended to existing members, respect for their lives, the integrity of their physical bodies, their well-being, their wills, their rights. If this is what Leopold meant—and it may be what he intended although he did favor culling animals and hunting them as a means of culling them when culling was "necessary"—it is an interesting, discussable position, one that has claims to be a new, ecological ethic. The community of the ethical is enlarged. Things, plants, animals, come to count, not simply in so far as their goods and states, are relevant to man's good, but as ends in themselves. It is to advance an ethic that is not human-centered, human-oriented, one that is not based on the concept of a person as the central concept of ethics. Equally, in Leopold's hands, this land ethic seems not to be thought of as a theistic, God-centered ethic. Yet considerable difficulties remain. Many things have no life, no wills, no well-being, and are logically incapable of possessing rights. The new ethic would seemingly dictate treating such constituents of nature as "ends in themselves."

Again it is hard to determine what this would come to. Perhaps it implies some sort of equality in nature, between the various constituents and between them and man. Whatever be meant, the practical implications of such a normative equality are anything but morally generous, morally enlightened, impressive. Rather they are reminiscent of the morally objectionable views and attitudes of primitive moralities. To treat a person, a dog, a fly, a flea, an ant, a malaria organism, a tree, a thistle, a stone, a grain of sand, a pool of water, an ocean, as equals, as ends in themselves, would be to act in a grossly immoral and morally insensitive way.

Fortunately, Leopold does make other statements that cast doubt on the correctness of such an interpretation, as when he observed: "An ethic to supplement and guide the economic relation to land presupposes the existence of some mental image of land as a biotic mechanism. We can be ethical only in relation to something we can see, feel, understand, love, or otherwise have faith in" (1966, 230). And: "That land is a community is the basic concept of ecology, but that land is to be loved and respected is an extension of ethics. That land yields a cultural harvest is a fact long known, but latterly often forgotten" (ibid., x). This cuts across the above interpretation of the land ethic and suggests some sort of personification of nature and its constituents, or a belief of a mystical kind in the personhood of things in nature. We cannot understand, love, have faith in natural phenomena as ordinarily understood unless we attribute personhood to them. No reasons are offered for doing this. Yet, if it is being suggested that we should understand, love, have faith in natural phenomena in some metaphorical sense, it is unclear how the metaphor is to be understood.

Alternatively, it is possible to interpret Leopold as proposing that all ecological phenomena be seen to be and be treated as members of the ethical community in some more basic sense, such as that in which men and sentient beings in preference utilitarianism are treated as such. Stones, mountains, forests, amoeba cannot have preferences. To attribute preferences to them is to speak metaphorically or in the very puzzling way adopted by Saint Thomas Aquinas, who attributed "inclinations," "appetites" to mere things in his account of the natural law.

There is a real problem in attributing a coherent meaning to Leopold's statements, one that exhibits his land ethic as representing a major advance in ethics rather than a retrogression to a morality of a kind held by various primitive peoples. The problem is all the greater if we remember him for what he was, a thoughtful, serious moralist who acknowledged the duties generally acknowledged by thoughtful men of good will, a realist, not a sentimentalist about nature, one who believed that man could and ought to intervene in nature for nature's good, that man could improve nature, help it and its members by acting in many caring ways. Because of these difficulties of interpretation, and because Leopold was no romantic idealist but one who accepted the propriety of man using nature to his ends, it is tempting to view him as simply another ecological moralist who, like Dubos and so many others, urges a caring attitude toward nature simply because that is in man's best interests. There are hints of such an approach, and it is true that he firmly believed that a caring attitude is in man's interests, but overall he obviously wished to say something much more radical. He did believe that a new, genuinely ecological ethic was necessary and possible and that he had made a substantial contribution towards formulating it.

R. Routley and V. Routley

Another attempt to develop a specifically ecological ethic and thereby to explain the basis and nature of our ecological duties has been made by R. Routley and V. Routley in various papers, the first important one being "Is There a Need For a New, An Environmental Ethic?" (Routley, 1973; see also V. Routley, 1975, and Mannison, McRobbie, and Routley, 1980, chap. 8). The Routleys note with approval Leopold's statement that morality should be extended to embrace the natural world, but they see the development of a specifically ecological ethic as being much more radically revisionary than did Leopold, as necessitating a new meta-ethic that does not base or explain, as do the currently fashionable meta-ethical theories, all valuation as human valuation, and as requiring the working out of a nonhuman-centered normative ethic that acknowledges nonhuman ecological and environmental values.

R. Routley introduces his demand for a new environmental ethic with a critique of what he describes as the liberal ethic. In fact, what he discusses and analyses is the liberal philosophy, what might be called the ethic of liberalism, which is quite distinct from the multifarious ethics of actual liberals. It is the liberalism of H. L. A. Hart, as expressed in his account of natural rights, and the liberalism emanating from J. S. Mill with which he is concerned. Routley's contention is that the ethics of liberalism is human-centered (human chauvinism), since it accords to persons the liberty to act as they please (including to destroy nature), provided they do not harm other humans. He neglects to note that the harm principle is usually extended to cover causing unnecessary suffering to many sentient animals, although not to all such. Routley is clearly right that this Millian philosophy of liberalism is heavily human-centered. He is wrong in identifying it with the morality and ethics of liberals. A key aspect of liberalism generally, including Hartian and Millian liberalism, is the insistence that the individual has the moral right to act immorally, provided that he or she does not harm others. That is what the debate about the enforcement of morals is all about. A liberal may argue that it is morally wrong to destroy forests, exterminate rare species, yet insist that the individual has the moral right to act in such immoral ways, provided that by so acting the individual does not harm others. Although, for the reasons set out earlier in this chapter, it is probable that such actions will often be harmful, and although the liberal has an elastic concept of harm, it is possible for a liberal, consistently as a liberal, to argue both that certain acts of vandalism of nature are morally wrong even though they do not harm others and that they ought not to be forbidden by the state.

The point of R. Routley's discussion of liberalism is to show that the ethical theories of their most serious opponents today are human-centered. His discussions would establish this only if liberals morally approve of all that they believe should be tolerated. Obviously this is not the case. Liberals insist that much that is immoral should be tolerated. However, as I have argued in many places, there are limits to this. To tolerate infanticide, cruelty to animals, corruption of the young, dueling, drunken driving, is to condone certain immoralities. There are obvious problems in determining when tolerance changes to condoning, condoning to accepting as morally right or as not morally wrong. To that extent Routley is right that the Millian political harm principle, in its toleration of the vandalism of nature that is not harmful to man, may express a human-centered ethic. In fact, Routley need not rest his case on an analysis of "the liberal ethic," nor need he have confined his charge of "human chauvinism" solely to liberals. He could have proceeded to examine the prevailing ethical theories and prevailing moralities. These could easily have been shown to be either God-centered, and via God to be human-centered, or to be very human-oriented, as, with the exception of duties to God and those concerning avoidance of certain kinds of animal suffering, they are heavily humanly-oriented, explaining most of our duties as relating to our fellow human beings. For many, the only recognized duties that do not fall under these characterizations are duties to promote good and lessen evil (these duties encompassing animal suffering and cruelty). However, many who acknowledge this general duty write in a human-centered qualification in respect of goods and evils other than animal suffering, that intrinsic goods all involve reference to experiencers, actual or possible, where the experiencers are usually thought of as persons. In this way, even ideal utilitarianism commonly becomes, like most other traditional and conventional ethical theories, heavily human-centered. Consider again the values most commonly stressed: respect for persons, justice, honesty, pleasure, happiness, human self-development, knowledge, rationality and rational belief, friendship, beauty and aesthetic excellence.

The negative part of the Routleys' argument, that traditional ethics and conventional morality are human-centered, is substantially well based. The ecological moral intuitions they seek to explain and justify find no justification in terms of such ethics. We should look now at the positive contentions of the Routleys. Here, because it is the general enterprise with which I am concerned, I shall not seek to distinguish the respective contributions of Richard and of Valerie Routley.

The Routleys insist that a new ethic is needed and make it clear that it must be one according to which the question of the proper response to our environment is not to be determined from a standpoint centered on the good of man. To this extent, the demand for a new ethic is one for simply a new normative ethic, one that could be developed within the framework of ideal utilitarianism. However, seemingly because they acknowledge that ideal utilitarianism does not provide an adequate framework for what they deem necessary— because ecological values are not to be explained simply as intrinsic goods, they seemingly being seen to involve intrinsic obligations comparable with those indicated by W. D. Ross in his principles of prima facie obligation—and because a meta-ethic with an account of valuation that is not one in terms of human attitudes, preferences, wishes, needs, is seen as also needed if an ecological ethic is securely to be grounded, they argue for the necessity of both a new meta- and a new normative ethic.

Although less dominating today than a decade ago, noncognitivism would seem still to be the dominant meta-ethic. It seeks to explain valuations as being logically tied to human pro-attitudes. If this meta-ethic is to be accepted, the Routleys are seeking what it is logically impossible to set out, a new ethic that is not human-centered in the very basic sense of making human attitudes, human reactions, the basis of valuation. My own view is that many conclusive arguments can be pressed against noncognitivist analyses, but this is still much disputed. Specific noncognitivist analyses are rejected, but there remains a widespread confidence that noncognitivism in the abstract is true and that it will eventually provide a sound analysis of ethical terms. If my view about the cognitivity of moral expressions is correct, this difficulty would not arise for the Routleys nor for other ecological ethicists. However, they do need to offer an alternative, non–human-centered meta-ethic. The range of alternative meta-ethics among those that have been developed to date is limited. They cannot embrace ethical relativism, nor the Humean ideal observer ethic, nor the Thomistic natural law ethic in terms of human nature and its natural end, nor naturalism including ethical subjectivism, nor Kantian rationalism. All make human nature or personhood or features thereof basic to ethics. Of the known, plausible, meta-ethical theories, the only one available to such ecological ethicists is the ethical realism of intuitionism. Acceptance of such a meta-ethic is normatively neutral. Hence, even if such a meta-ethic were to be shown to be the true one, as I believe it can be, the ecological ethicist who adopted it would still be left with the task of associating it with a non–human-centered normative ethic, one that did not locate all values in human capacities, states, goods, but accorded to environmental phenomena value in their own right, or to respect for them obligatoriness in its own right. (See McCloskey, 1969.) The Routleys decline to seek a solution to their problem in terms of intuitionism. That is not to say that it cannot and ought not to be accepted by an ecological ethicist. On the contrary it would seem to be the most promising line of approach for basing an ecological normative ethic.

Instead, the Routleys, in an obscure discussion, seem to see their problem as that of explaining valuation in terms of valuers or valuing but not in terms of human or other determinate kinds of valuers. If values are tied to valuers or to valuing, it is hard to see how the kind of ecological moral judgments that it is sought to explain and justify can so be explained and justified. The ecological ethicist wishes to ascribe value to nature, both that existing prior to the evolution of man and that existing after the extinction of *Homo sapiens;* and he wishes to argue that there are basic, intrinsic obligations to respect nature either by virtue of these valuations or as a matter of ultimate principles of morality, where the principles in no way relate to goods or values to be enjoyed by man. An alternative solution of basing valuing on valuers but not on actual valuers, along the lines of adopting a phenomenalist account of values, a permanent possibility of valuing, is exposed to difficulties parallel to those encountered by phenomenalism itself.

Although they do not say as much, and although they do not use this terminology, the Routleys appear to base their normative ethic, which is nowhere set out and developed in a systematic, detailed form, on the basis of intuitive insights into what is intrinsically valuable and what is intrinsically obligatory, where what is intrinsically obligatory need not be tied to intrinsic value. Species and wilderness seem to be attributed intrinsic value, but not necessarily intrinsic goodness. (The difficulties in the way of attributing goodness were set out earlier in this chapter.) This raises the question of the nature of the value they are believed to possess. Acts of violating, vandalizing nature, wood-chipping a rain forest, strip-mining a wilderness area, exterminating a species, seem to be seen as intrinsically wrong quite independently of the destruction of the intrinsically valuable they involve. However, the exact nature of the normative theory, whether it is one solely in terms of intrinsic value, or intrinsic value and the intrinsically obligatory, is not made clear. Nor are the relevant values or principles explained or discussed. Clearly, the ecological normative ethicist, if he is to offer a serious ethical theory, must state it clearly and fully so that it can be examined and assessed. That still remains to be done.

Meanwhile, it is possible to note some of the difficulties this kind of ethic encounters. It will be confronted as fully by the problems noted earlier concerning the preservation of disease organisms, rabies, malaria, and parasites such as lice, fleas, as is an ecological ideal utilitarianism. There seems to be no intrinsic value in the existence of such species and organisms. There appears to be no intrinsic duty to protect and preserve these species that must be overriden for us to be justified in eradicating the species and their members, other than where they are beautiful.

Are there other approaches that may be followed in seeking to formulate and defend a new, ecological ethic? A possible approach might seem to consist in seeking to rest such an ethic on a new, distinct, yet ethically basic sense of "good." At first sight this appears promising. In fact it amounts to suggesting only that a new meta-ethic may be devised in terms of a new analysis of "good." The difficulties in the way of seeking to develop a new meta-ethic by way of definitions of basic ethical terms are too well known to need rehearsal here. Similarly, an approach suggesting that there are other, basic moral words, not acknowledged by conventional ethical theories yet used to express ecological value judgments, value expressions such as *magnificent, precious, sublime, arousing wonder, awe,* and so forth amount to suggestions that our existing meta-ethics are very inadequate in their scope, that they need to be supplemented by reference to new value terms and analyses of the concepts to which these terms relate. Against this, it may be argued, as it would be by exponents of prevailing meta-ethical theories, that insofar as they have a significant meaning as value terms, rather than as mere expressions of emotions, the valuation they express must ultimately be explained in terms of the more basic value concepts, *good, right, ought, duty, beautiful.*

Why it nonetheless is important to take seriously the possibility of an ecological normative ethic, in spite of all the difficulties that have been noted, is that the considerations outlined in this chapter as bearing on the wrongness of not preserving but needlessly, wantonly destroying species and wilderness, imply that in needlessly destroying a useless, ugly species such as the Rottnest quokka (a small wallaby) we should at worst be acting with very slight imprudence and somewhat reducing the scope for human knowledge and pleasure. Yet the wrongness of the deliberate, unnecessary extermination of this species seems to be vastly greater than that. We must either reject these intuitions or develop an ecological normative ethic.

To sum up the case for preservation: The case for preserving natural phenomena, wilderness, species is defined, explained, and circumscribed by reference to our duties to preserve the valuable, the intrinsic good, and so to act as not to harm but to help the living and those of future generations, where these duties genuinely hold. Hence it is determined by the duties of promotion of good, prevention of evil, respect for persons, those now living and persons yet to be born, justice, the rights of the living and the rights those who will come to exist will possess when they come into being.

6

Animal Rights? How They Would Bear on Ecological Ethics

There has been much philosophical writing concerning moral rights during the past fifteen years. One of the most discussed topics relates to who or what can be a possessor of a moral right, the concern of most writers being with the question of whether animals, and not simply human beings, can possess rights. Initially, it was those who wished to develop more powerful arguments in suport of condemnation and even outlawing of cruelty to animals who explored this approach. More recently, they have been joined by environmentalists who hope to find in theories of moral rights new and powerful arguments for respect for nature and its constituent animals and plants.

Much of the writing confuses the case for according legal rights to nonhumans, to animals, plants, forests, mountains, and ascribing moral rights to such things. There is clearly a great difference between arguing that animals, plants, trees, mountains possess moral rights and that it is morally desirable and expedient to accord them legal rights. This confusion has detracted a great deal from the force of many of the relevant arguments. Clearly it is nonsensical to suggest that inanimate things, buildings, works of art, mountains, possess moral rights. It is not absurd, if they are valued and valuable, and if they are threatened, to argue that they ought to be accorded legal rights, and that certain persons, or even all parents with an interest in the future, or all citizens having an interest in these things at the present time, be given standing, and thereby be empowered to take legal action to protect them by protecting their legal rights. The idea is that if rain forests, an endangered species, or the like were to be accorded legal rights to life, integrity, freedom from harm, and if certain persons were given standing so that they could act as guardians, being legally empowered to seek to protect the legal rights of the natural phenomena, the natural phenomena might be safer from wrongful destruction than they now are. Those who seek to harm them could be opposed by court action.

It may well be that this could be an effective, desirable way of providing protection for what is valued, although it is an approach that is not without its problems. Firstly there are obvious difficulties in determining what ought to be protected in this way, who should be given standing to act as "guardians," and the degree of discretion the courts should have to determine when action that is harmful to the natural phenomenon is justified. Extending the class of natural phenomena so protected, for example, by recognizing them as legal persons, and opening the class of those given standing, could lead to important, essential projects relating to energy production, irrigation, mining being seriously delayed and thwarted by court action and the costs of delays, which would usually include not simply the legal costs, but interest on capital and equipment tied up while the court's decision is reached. On the other hand, to restrict these classes is to risk there being inadequate protection. Other difficulties arise because such rights could not possibly be construed as absolute legal rights. Yet to accord to courts the power to determine when they are legally overriding is to put into the hands of judges who are not accountable to the community power that is more appropriately exercised by the legislature. C. D. Stone's thoughtful discussion of this kind of approach in *Should Trees Have Standing: Towards Legal Rights for Natural Objects* brings out some of the advantages, and by way of incidental comment some of the dangers, of adopting this course of action.

Many who are concerned about the welfare of animals, and who have sought a charter of (legal) rights for animals, have claimed and even believed themselves to be arguing that animals possess moral rights, when in fact their arguments simply relate to the case for legally protecting animals from abuses by according animals legal rights. This is a possible position to adopt. It is quite distinct from the view that animals do in fact possess moral rights. Since the most promising, although not the only, way of arguing for legal rights for animals is that they possess moral rights, we might usefully first consider the question: Who or what may be the possessor of a moral right?

There seems to be no problem at all about ascribing rights to persons. To my knowledge, the only philosophers who question whether rational adult persons can and do possess rights are those who reject the concept of a moral right, and hence the whole theory of moral rights. All who talk about moral rights as things that are of moral importance accept that human persons are paradigm cases of possessors of rights. Doubts arise in respect of very mentally defective human beings, "human vegetables," who are incapable of self-consciousness, even of consciousness. Some, including myself, argue that such beings cannot possess moral rights, that they have no selves, no personalities, actually or potentially, and cannot claim or come to claim any rights they might be said to possess. Others contend, against this, that since they are human, they must possess moral rights that can be claimed on their behalf by others who act as their guardians or representatives. This has led others again to maintain that, if such human beings can possess moral rights, then animals with similar attributes can also possess moral rights. Hence it is further argued that the class of beings that may possess moral rights is not the class of persons, nor that of human beings, but that of sentient beings. Such *ad hominem* argumentation is obviously of very limited value.

It may be useful to sketch in how the argument has proceeded in recent years. A good deal of writing about animal rights, for example that of T. Regan, R. G. Frey, J. Feinberg, has taken as one of its starting points the contention that only beings that can properly be said to possess interests can possess rights. (See Nelson, 1956, and McCloskey, 1965.) I had argued that interests could not properly be attributed to animals, only to human beings, to persons and to potential persons, my thought being that when we analyze what constitutes a person's interests we are not thinking of what he desires, or even of what he needs, but what is for his good, where his good is determined by what he is, by the role he has adopted, the aspirations he has, such that it may be in his interests to do what he does not desire to do, and even to forgo satisfaction of various of his needs. I argued for the importance of this on grounds that included the thought that it is empty and meaningless to ascribe rights to beings who could not claim them and who could not be represented, for whom guardians could not act to claim and defend them in respect of their enjoyment and exercise of their rights. If rights are dependent on interests, then if interests do not hold rights equally cannot hold, since they would then not admit of representation or protection by a representative or guardian.

T. Regan, in numerous discussions (see for example, Regan, 1975, 1976a, 1976b, 1977, 1979), and J. Feinberg in his imaginative article, "The Rights of Animals and Unborn Generations" (1974), accept the point concerning the relevance of interests and the possibility of representation to rights but reject my claim that sentient animals do not have interests. Feinberg argues that where the notion of acting for a being's *sake* has meaning, then the being has interests, such that we can meaningfully speak of acting for the sake of a sentient being, that a sentient being therefore has interests and rights that can be protected by a representative or guardian, that its rights and interests coincide, and hence that it can have rights that it cannot itself safeguard, but that can be safeguarded on its behalf by representatives or guardians. R. G. Frey in his many papers and in *Interests and Rights* directs his arguments against Regan's and Feinberg's rejection of conclusions such as those I support; he offers new arguments and contends that it has not been shown that animals can have interests in an acceptable sense of *interests*. Frey's argument is a complex one, being both *ad hominem* against various of Regan's arguments and of a general character relating to the interconnections between interests, desires, and beliefs. To report it accurately here would take more space than is available. Discussion of it now impresses me as unnecessary because this whole approach to the question of animal rights by way of the question as to whether animals can and do possess interests seems mistaken, misdirected, and misconceived, for the reasons I developed in "Moral Rights and Animals" (1979). There were two serious errors in my original discussion. I was mistaken in claiming that, among all organisms, only human beings can possess interests. In fact, any organism, a plant, a tree, a frog, a cat, or a human being, can and does possess interests. Of course, what makes for the interests of a shrub, a tree, a frog, a person, differs. It was this that misled me into believing that because what contributes to the interests of a plant is different from what contributes to the interests of a person that a plant cannot have interests. Yet, clearly, it is against the interests of most plants to deprive them of water, to cut their roots, and so on. So, too, it is against the frog's interests to drain the pond in which it lives. Animals and plants may have interests. If this is so, interests can no longer provide a basis for marking off possible possessors of rights from those beings and things that cannot possess rights. Some possessors of interests logically cannot possess moral rights. (In passing, it is worth noting that acknowledgment that plants and lower animals possess interests underlines the indefensibility of Singer's principle of equal consideration of interests. See Singer, 1975.) Rights and interests are distinct, and it is a very serious error to confuse that which is in a being's or thing's interests and that to which the being or thing has a right. The being or thing may be incapable of possessing moral rights; and where the being can do so, it may have no right to pursue its interests.

What are in one's interests may be evil, and such that one may have no moral right at all to the satisfaction of one's interests. It may be in one's interests that a wealthy uncle be killed, since he has willed you all his property. Yet you, in virtue of your interests, have no moral right to kill him, not even a *prima facie* right to do so. Similarly, it may be in your interests to overwork your horse cruelly, yet you have no moral right to do so unless your interests are morally worthwhile. Or, to take the kinds of cases cited by G. Harman in *The Nature of Morality* to illustrate another point, it may maximize the satisfaction of interests to abduct a healthy person, kill him, first taking his rare blood to save the life of one person, transplanting his heart and kidneys to save three others, and his corneas to restore the sight of another. Yet it would be grossly immoral so to act. The parties concerned would have no moral right to engage in or condone such a course of action. This is because there is a vast gap between interests and rights.

This gap can be illustrated another way. To represent a person and protect his interests is very different from representing and protecting his moral rights. Paternalists, as liberals frequently note, infringe upon people's rights for the sake of protecting their interests. The liberal demands the right to exercise his moral rights even in ways that are damaging to his interests. The following example illustrates the divergence of rights and interests. A wealthy man, Jones, who each year gives generously to a worthy charity that has come to be dependent on his donation, has a stroke and falls into a coma. The donation comes to be urgently needed by the charity. Smith has a power of attorney to act for Jones. What should he do? If he asks himself, What course of conduct is in Jones's best interests?, he may well decide to give no donation. If, on the other hand, he asked himself, How would Jones have exercised his moral rights, and how would he wish me to exercise them for him?, it is likely he would decide to give a donation comparable with those Jones had previously given.

This is very important to the issue of whether animals can possess moral rights. Moral rights typically can be waived, forgone, insisted upon, exercised in or contrary to one's interests. For possession of a moral right to be meaningful, the possessor or his/her representative must be able to claim it, exercise it, or the like. Consider here a legal right that no one could legally exercise, and which no one could claim or exercise on the possessor's behalf. It would be empty talk to claim that any legal right existed under such circumstances. The same thing holds of moral rights. Yet this is the situation that would prevail if animals were to be attributed rights. We can plausibly claim to have some idea of what is in the interests of an animal. We can have no knowledge concerning how it would exercise its moral rights, if it possessed any. This is because it has no capacity to exercise them, while a capacity to do so, actually or potentially, is presupposed by talk about moral rights.

In brief, a consideration of why paradigm cases of possessors of rights, persons, rational, morally autonomous beings, are such, and why the exercising of rights by the possessor of the right or a representative is so central to the concept of a right brings out that it is capacity for moral autonomy, for moral self-direction and self-determination, that is basic to the possibility of possessing a right. Consider how we should respond if it were to be determined that a whale or a dolphin possessed moral autonomy.

To show that animals do not, cannot possess moral rights is distinct from claiming that we have no duties in respect of animals. Clearly we have many important duties in the treatment and care of animals. The claim about rights is simply that the duties we have are not and cannot be based on the possession of rights by animals.

Environmental Implications of the Possession of Moral Rights by Animals, if Animals, Some or All, Possessed Rights

It is worthwhile considering what would follow, environmentally, if animals did in fact possess moral rights, as both many animal rightists and many ecological moralists wrongly believe that the case for preservation and more generally for respect for nature would be strengthened by showing that animals possess rights and by bringing about respect for those rights. Far from contributing to a lessening of the problems commonly referred to as "the ecological crisis," the possession of moral rights by animals, and the respecting thereof, would dictate further encroachments on wilderness and hence further harm and endanger many animal and plant species. Respect for animal rights would make more serious many of the other problems that are said now confront us. Consider what rights sentient animals might be supposed to possess. Rights that are most commonly claimed to be held are the rights to life, freedom from suffering, and liberty ("animal liberation"). Few seem prepared to contend that animals have a right to breed at will, but seemingly, if that right could be shown to hold for humans, it might also be shown to hold for those animals which possessed other moral rights.

As with human rights, animal rights conflict with one another. The rights of a cat would come into conflict with those of other cats, and the rights of cats would conflict with those of mice, birds, other carnivorous animals, and so on. They would also conflict with the rights of persons. Man, as the only possible guardian of animals' rights, would have to resolve these conflicts and protect the rights that were the operative ones, and thwart those animals that wrongly sought to invade the rights of other animals and of humans. The mind boggles at the difficulty and complexity of this rights calculus. An easier calculus may be suggested as being the appropriate one, namely, that man simply protect animals from human violations of animal rights. This is to ascribe to animals lesser rights than those ascribed to persons. In terms of either calculus, man would be faced by conflicts between the rights of domesticated farm animals and those of wild animals in respect of the use of land. The rights of domestic pets and those of various farm animals and wild animals would also conflict, since the former feed on the flesh of the latter. Man would either have to kill or let die out by preventing them breeding, domestic pets, cats and dogs, or kill farm and wild animals to feed them. The rights of disease-carrying pests such as rats and mice would conflict with those of humans to life and health, and with those of domestic pets and of various farm animals. If animal rights included the rights to liberty and to freedom from restrictions in breeding, we should respect the rights of domesticated animals, farm animals and pets, by letting them run freely and breed at will. The ecological implications of doing that are obvious and immense. If these animals do not possess such rights, we could in the long term resolve many conflicts by not allowing pets and farm animals to breed, thereby ridding the earth of all or most of them. (A small sample might be retained in naturally confined locations such as small islands.) This course of action would also have a vast ecological impact on the ecosystems of both cities and rural areas. To allow cattle, sheep, goats, pigs, fowls, dogs, cats, and the like to roam freely would be to change the ecosystem in ways very disadvantageous to man and very damaging to nature, to wilderness, and to rare species of plants and animals, the more so as various of these animals would reach plague proportions. To eliminate domestic pets from cities would be to bring about rat and mouse plagues, with consequent increase in disease and harm to human beings; to eliminate farm animals would in itself be to change drastically the ecosystems of rural areas. To replace these animals' grazing lands with cereal crops would be to change the ecosystems even more radically.

Those who stress animal rights see as the major ecologically desirable implication of their view that it dictates the acceptance of vegetarianism as a moral position and practice. However, if we are morally concerned about the environment, we need to look very closely at the effects on his environment of man opting for moral vegetarianism. A conclusion that is often drawn is that vegetarianism would lead to an early improvement in the world food situation, since there is now a considerable wastage of food in the conversion of vegetable protein into meat. However, the immediate effect of the adoption of moral vegetarianism would be a loss of a major source of food, meat. It would only be after farm animals are bred out of existence that the vast amounts of vegetable food would become available to man. Then it could be as if the world food situation were to be improved by millions of new hectares of land becoming available for production. This is qualified by a number of considerations. First, the transition period could be extremely difficult. Second, there would be a loss of many food-producing hectares of land and of water. Much land now used for grazing, such as much of Australia's Northern Territory, would become unproductive or less productive of food, because it is unsuitable for agriculture. If fish and other aquatic animals possess moral rights and are rendered unavailable as food by a moral vegetarianism, there would be a total loss of that food source, it presently not being possible for man to use the oceans, seas, rivers to produce vegetable food in place of this important animal food. Further, consistent action on the basis of moral vegetarianism with its concern for animal rights and/or animal welfare would allow great waste to occur as a result of not checking, or not checking by killing or harming, pests such as rabbits, rats, mice, birds of various kinds, which damage crops and food. There would also be the loss of food from meat that can be produced using otherwise useless vegetable matter with the intervention of microorganisms. Thus, while there would come with moral vegetarianism an early although not immediate improvement in the world food situation, it would be a much more qualified improvement than moral vegetarians would have us believe to be likely.

The ecological costs, especially in the longer term, could be considerable. The switch to a vegetarian diet would be at the cost of converting some or all arable grazing lands to cereal and other crops. This would have vast ecological implications in respect of pesticide and fertilizer pollution, changes to ecosystems through the change in land use, and the building of dams and the like, that would go hand in hand with this change of use, and the damage to the land itself that may well occur in the process. Further, if Malthus were to be correct in one of his claims, there could be pressure to use all the land now available for agriculture for cultivation of crops. Grazing although not overgrazing is far less damaging to many environments than is the cultivation of crops. Hence, while the switch to moral vegetarianism and practice can be made to be beneficial in increasing the amount of food available, it could prove very damaging to the environment in many ways such that in the long term less food may result than would have occurred had meat-eating remained acceptable, unless the switch to a vegetarian regimen were very carefully thought out and controlled.

Thus, contrary to the beliefs of many environmentalists and exponents of the view that some animals possess moral rights, that the two movements, animal rights on the one hand and conservation and preservation on the other, are complementary, mutually supportive positions, they in fact are fundamentally opposed and are such as to be calculated to lead to many major clashes. R. G. Frey in *Interests and Rights*, following J. Rodman's arguments in "The Liberation of Nature?," suggests that the animal rights thesis simply extends the range of those beings accorded the right to ravage nature. In itself, it provides no grounds for protecting nature from being ravaged by the privileged group of holders of moral rights. Rodman had argued to this end observing: "In the end, Singer achieves 'an expansion of our moral horizons' just far enough to include most animals, with special attention to those categories of animals most appropriate for defining the human condition in the years ahead. The rest of nature is left in a state of thinghood, having no intrinsic worth, acquiring instrumental value only as resources for the well-being of an elite of sentient beings" (Rodman, 1977, 19).

In summary, only those animals that are persons can possess moral rights. From the ecological point of view, this is fortunate. Were it to be the case that sentient animals possessed moral rights similar to those possessed by human beings, respect for them would greatly complicate and aggravate the very considerable environmental problems that already confront us.

7

Resource Depletion, Science, Technology, Pollution, and Population

Our human-centered, ecologically colored obligations relate to those living today and to those yet to be born. Issues relate to our use of resources, whether we are using them as we ought, with due regard for the rights and well-being of those now living and the well-being of those yet to be born; to our use of scientific knowledge and technology, and the powers they confer on us; to our duties and responsibilities in respect of reducing pollution; and our obligations in respect of population. Other non–human-centered obligations, some of them very important, also arise out of or in connection with our depletion of resources, our use of scientific knowledge and technology, our creation of pollution, and population growth. These obligations relate to the avoidance of unnecessary, avoidable animal suffering and the destruction of the beautiful, where the beautiful object may be a natural object or a human artifact. However, the major moral problems in these areas are those that relate to human persons. These will be the concern of this chapter.

Human Moral Rights and "the Ecological Crisis"

Human rights figure prominently in accounts of the nature of the ecological crisis, and also in discussions of possible solutions thereto. Some writers approach the problems by seeking solutions within a framework that admits of respect for the basic human rights. Others argue that the crisis is so grave that there is no human right that cannot legitimately be overriden, indeed, that in survival situations of the kind mankind faces, moral rights and moral considerations lose all relevance and validity. This latter view will be rejected here, and an attempt will be made to indicate how human rights determine what are morally acceptable solutions to the problems known as "the ecological crisis."

The basic human rights are the rights to life, health, bodily integrity, respect as a person and all that that implies, moral autonomy and integrity, self-development and education as dictated by that and other rights, and knowledge and true belief. Rights that many claim to be basic moral rights, and which are central to discussions of possible solutions of "the ecological crisis," include the right to marry, reproduce, own private property, and have privacy. The view to be advanced here is that the right to liberty is an aspect of the right to respect as a person, and that it is with qualifications, dictated by the rights to self-development, knowledge, and true belief, as well; that the property rules and rights should largely be determined by utility in the broadest sense of that term, while the right to privacy is a conditional right, that holds only when dictated by considerations of justice and respect for the basic rights of persons. Family rights raise some of the most difficult problems. Insofar as there are such rights, they hold simply as aspects of the basic rights to respect as a person and to self-development.

These moral rights do not simply figure in ecological moral and political discussions of the nature of the problems that confront us. They determine how these problems arise and are to be formulated. It is because human persons possess the rights to life, health, bodily integrity, self-development, that there is rightly concern about conservation, and much of the problem concerning the pros and cons of conservation relate to the moral rights of those now living vis-à-vis the moral rights those yet to be born will come to possess. The latter do not now possesses rights since beings who do not exist cannot possess rights. Nonetheless, morally we must take account as best we can of the rights that these unknown persons who will come to exist will possess when they come into being. Much of the concern about pollution relates to the fact that the more serious pollution threatens life and health, and in so doing, causes avoidable suffering. For vast numbers of people, overpopulation would result in lack of enjoyment of their rights to life, health, self-development, and even to lack of respect as persons. Rights also figure in accounts of possible solutions to environmental problems. Some proposed solutions morally shock because they involve total disregard of basic human rights of vast millions of persons. Consider here the proposals of the moralists who see letting die, letting suffer, leaving without medical aid, coercing and even forcing the consciences of persons, as not constituting illicit violations of human rights. Many proposals directed at the problem of possible overpopulation raise questions as to their moral availability in view of the claim that we possess such moral rights as those noted above. Yet, at the same time, many of these proposed solutions spring from a concern that all who are born be able to be secured full enjoyment of their rights to life, self-development, health, and to respect as persons. Proposals relating to reducing pollution and to checking resource depletion, securing preservation of wilderness and of endangered species, encounter criticisms based on claims that they would lead to violations of legitimate rights to private property. If the views to be defended here are sound, the latter criticisms need to be assessed in the light of the status and basis of private property rights as very qualified, derivative, and socially and economically relative rights.

The Right to Life

Although the right to life is very evidently basic and important, it is a right that has received relatively little attention from philosophers. What discussions there are are not impressive. This is accountable in part by the fact that for a long time the right to life was given a theistic justification, as by John Locke, in terms of our being God's property, and hence such that no one, including ourselves, can have the right to take our lives, while all have the duty to help preserve human life. Such a theistic account will not do. It implies that there is no intrinsic, human right to life, that it is God, not human persons, who possesses rights. Similarly, the Thomistic account in terms of the natural law, in spite of the great importance given to this right by Thomists as an absolute right, is completely unsatisfactory. We are said to have an inclination or appetite in common with substances to preserve ourselves, and from this inclination, the right and duty to preserve ourselves are said to follow. A more promising defense is that, in terms of man's nature as a morally autonomous, rational, imaginative being, possessed of a capacity to experience, feel, and recreate emotional and intellectual experiences, such that when we reflect on what is involved in the existence of an autonomous being we see that it has a right to respect, killing is intrinsically although not necessarily absolutely wrong as violating that which morally ought to be respected. Thus the right to life is to be seen as self-evidently resting on man's nature as a person, that is, as a morally autonomous being, and on the respect due to the worth of this kind of existence. To destroy such a being is to act in a way comparable with willfully destroying a valuable work of art. It rests also on respect for and acknowledgment of the fact that possession of rational autonomy gives the person who possesses it rights over his/her own existence, he/she having the capacity and hence the right to determine what happens to him/her. Respect for a person involves respect for that person's will. How can another have the right to override the will of a rationally autonomous being? The right that follows from these considerations is not simply a negative right, a right not to be killed, but a right of recipience, a right to be aided in being kept alive. If existence as a person has such worth, clearly persons have a right not to be let die. These issues are discussed at greater length in my "Right to Life" (McCloskey, 1975).

This kind of defense of the right to life through the nature of man as a person is not uncommon today. It is nonetheless not without its difficulties. An obvious difficulty relates to the fact that infants are not morally autonomous, yet to kill or let die an infant would seem very evidently to be violating a right to life. Here allusion needs to be made to their potentiality to be persons, that it is persons and potential persons that possess the right to life. To kill an infant does not violate its actual will, only what would be his/her will, if the infant could will. It is also to destroy something of immense worth, a potential person, where, besides the core characteristic of personhood, and hence of moral autonomy, there is also the capacity to imagine, feel, recreate intellectual and emotional experiences, and the like. If this move is sound, it follows, in the absence of other reasons to the contrary, that the fetus, from the moment of fertilization, has a right to life by virtue of being a potential human person. There is a lot of loose, philosophically careless use of the word *potential* such that sperm and ovum before fertilization are often described as potential human beings, and according to which, mere possible human beings are said to be potential human beings. That is not a morally significant sense of *potential*. The morally relevant sense is that in which we speak of something as a potential X when, inherent in the thing itself, is the natural development into an X. It is in this sense that the fetus is a potential person.

Given this sense, difficulties arise in maintaining that all potential persons have the same rights as do persons. There is general agreement that a day-old infant is a potential person, and that it is intrinsically wrong to kill such a being. Yet there seems to be no clear way of justifying sharply distinguishing a fetus in the later stages of its development from the day-old infant, and the less developed fetus from the more developed one. Arguments that stress viability (which now is very evidently technologically relative), or which suggest that the fetus in its early development is simply a potential organism, not a potential person, or which represent the fetus as simply an appendage of the woman's body with which she can do as she will, fail to answer the fact that, except for defective, damaged fetuses, in any serious, hard-core sense of *potential* the fetus at all stages of its development is a potential person. The relevant consideration overlooked by such arguments is that while the potentiality of a being is morally important, so too is its actuality. It is morally wrong to treat a potential person, an infant, simply as a person; so too with the fetus, more evidently at the early stages of its development. Its actuality as the kind of being it is is also morally relevant. This is because a potential X is not an X, and what holds of a potential X need not hold of an X. This is evident enough in respect of the potential frog, the tadpole. Frogs and tadpoles differ greatly, and killing a frog is different from killing a tadpole, just as killing a fertilized hen's egg is different from killing a hen, and the killing of a human egg is very different from killing an adult person, the more so as the former, although a potential X, is for a time at least not a separate, distinct organism in its own right. What follows from this, I suggest, is that qua its potentiality the life of a fetus should be respected, but qua its actuality as a developing entity, this is less evidently true. Its potentiality as a person tells against the permissibility of abortion. Its actuality at the early stages of its development may tell in favor of the permissibility of abortion. The potentiality of the fetus as a person, while not making abortion and infanticide morally equivalent, does make it morally inappropriate for persons to plan to use abortion as a method of birth control, given the available alternatives.

All this is qualified by the fact that the right to life, like the other basic rights, is a prima facie right, one that may be overridden by telling moral considerations. Thus if the child to be born cannot adequately be fed or cared for, if it will be born to eke out a wretched, degrading, brief existence, that fact would constitute grounds for the moral permissibility of abortion. To be shown to be morally acceptable as a method of population control, it would need to be shown both that more morally acceptable methods of population control are not morally available (as we shall see, they are) and that its use is justified as being necessary to avoid the wretchedness and misery that come with overpopulation.

There has been a lot of thoughtful, impressive writing about the morality of abortion in recent years. (See for example Feinberg, 1973, 1974.) None of it is philosophically convincing in a decisive way. One of the weakest arguments is the common feminist argument that proceeds from the claim that women have the moral right to control over their bodies, where the fetus is represented as being simply a part of a woman's body. The fetus is not simply a part of a woman's body. In any case, it is unclear whether we can meaningfully, let alone rightly, be said to "own" our own bodies, and have the right to do with them what we will. Such a contention harkens back to a very philosophically primitive notion of property and of property rights as free of all responsibilities. If sound, at most it would establish that the woman, no one else, had the moral right to decide whether the fetus be killed. Compulsory abortion as a method of population control would, on such an account, violate the woman's right to control over her body. Thus, both this feminist argument about abortion and the more conservative arguments entail that abortion is morally unacceptable as a method of coercively controlling population growth.

The Right to Health

A right to health is obviously derivative from the right to life. However, if it is to be derived only from the right to life, it would be a much lesser right than is commonly claimed to hold of persons. Much ill health, mental and physical, does not endanger life. The full right to life is tied up with the rights to bodily integrity and the right to respect as a person.

The right to bodily integrity is obviously closely related to the right to health. In defending such a right, it is easier to argue from the lack of rights of others to injure, infect, deliberately disease, mutilate another. However, on the positive side, there is the right to determine what happens to one's self and one's body. It is yours in the sense of being part of you as a person, and you have the right to care for it as part of your person. Many claim—as do feminists in their arguments for the woman's right to abortion, libertarians, and Lockeans in their formulation of the labor argument for the right to private property— that we *own* our own bodies. This is seen as the origin of our other property rights. This would seem to imply that we have the right to dispose of our bodies at will. There are obvious difficulties with this kind of notion of private property rights in one's own body; and, in any case, if such ownership rights were to be meaningful, the rights could well be ones of a stewardship, or such as to permit the use but not the harming or mutilating of what we own. How this issue is to be resolved bears importantly on whether sterilization constitutes a violation of a basic human right, and hence on whether it is morally available either as a voluntary or coercively applied method of population control.

The right to health (and the right to bodily integrity) are thought of as rights of recipience. They are so, insofar as health and bodily integrity relate to us as persons and to our well-being as persons. It is the thought that these rights are rights of recipience that provides the philosophical and moral basis for the welfare state in which everyone has access to that medical and hospital care which is necessary to maintain good health, mental and physical.

Insofar as there are moral rights to life, health, bodily integrity, the proposal that the wealthy nations of the world, by continuing their restrictions on immigration, their trade policies, tariffs, market agreements, and lack of benevolence starve peoples into accepting those population measures that they (the wealthy states) favor, whether or not they are morally unacceptable to those upon whom they seek to impose them, is a grave violation of these human rights, as well as of other rights yet to be discussed, most notably the right to respect as persons.

The Right to Respect as a Person: Moral Autonomy and Integrity

The rights of persons to autonomy, more especially to moral autonomy, appear to be the most self-evident rights, even though, like other rights, they are not always absolute rights. People wish to do strange and evil things, even from the moral conviction that they ought to do so. To deny respect for a person's moral autonomy and integrity without good reason is to treat that person as a thing, and to deny his/her personhood. To force a person to act against his/her sincere, carefully thought-out moral beliefs, is to deny that person his/her autonomy in respect of the most important aspect of his/her life. This may, on occasion, be necessary, as with conscientious murderers. However, one needs grave moral reasons for thwarting a person's actions. One needs even graver moral reasons for thwarting a person's *moral* actions; and even graver moral reasons for forcing a person to do what that person believes it is morally wrong to do.

This right is an important part of the basic data of which account must be taken in proposed solutions to environmental problems. Many proposals for controlling population growth demand that people be forced to do or accept what they morally believe they ought not to do or accept. Many measures directed at preservation and even conservation will prevent persons from doing what they believe they ought to do. If this right is to be fully respected, voluntary methods will generally have to be relied upon to solve the ecological problems. Yet in many areas these are very evidently inadequate. To override this right legitimately we need to be justified in terms of the facts, that this overriding is necessary and justified in terms of the evils prevented. In nonecological situations, we usually believe ourselves to be justified in forcing a person to do what he believes he ought not to do, only very rarely and where the justification for doing so is very strong. Consider conscientious objection in war, the treatment of those parents who believe it to be wrong to have blood transfusions, medical treatment, for themselves or their children. We rightly seek to avoid clashes in which the conscience of the person is forced, and rightly believe that the law is defective and the whole moral situation unsatisfactory as things now stand in respect of conscientious objection in war. Yet many environmentalists are proposing that strict Catholics and the many others who morally object to the practices be forced to accept sterilization, to use artificial methods of contraception, to have abortions that are seen as being as evil as child murders, and so on. In other areas of morality, we are hard-pressed to find a case in which we believe it to be right to force thoughtful, rational, sincere persons to do what they deeply believe they ought not to do. Nonetheless, ecological moralists, most notably those of the United States, propose that the conscience of hundreds of millions of persons should be forced. The message of Roger Williams that the forcing of the conscience is a spiritual rape, that a forcing of a person's mind and soul is far worse than a physical rape, being an abomination of the worst kind, has been lost sight of by many of his twentieth-century countrymen (Williams, 1644).

The right to self-development and the right to education and other rights that come with a right to self-development are based on the nature of the self-development of persons. It is self-development, not any self-activity, that is the object of that right. That is why the right to liberty that follows from the right to self-development is a right to a lesser liberty than that which follows from the right to respect as a person. The right to self-development is a right to full development as a human person and carries with it a right to access to what is necessary for this. Any solution to the ecological crisis, if it is to be morally acceptable, must be one that allows all persons to enjoy this right.

The right to knowledge and true belief is a very important basic right, and one which follows from the rights to respect as a person and to self-development. It carries with it a right to a large measure of freedom, more especially the freedom to inquire and to report the results of one's inquiries, and more generally the freedom of expression and discussion. Acknowledgment of this right provides a check on the kinds of totalitarian measures a state may adopt to solve ecological problems. Thus conditioning, brainwashing to inculcate "useful" attitudes, will be incompatible with this right, as well as with the right to respect as a person. On the positive side, recognition of this right would do much to secure an open society.

Family Rights and Rights to Reproduce

Most persons in liberal societies cite the rights to marry and determine whether one has children, and if so, how many children, as among the most evident and basic rights that persons possess. Perhaps it is for this reason that they are relatively little explicitly discussed in writings concerning rights outside those of the Thomistic, Catholic traditions; they are simply accepted as a self-evident rights. I suggest that insofar as there are rights to marry and reproduce, they fall under two basic rights: the right to respect as a person and the right to liberty as a dictate of that right, and the right to self-development. Among the most important decisions, the most important exercises of personal autonomy, are those of deciding whether to marry, whom to marry, whether to have children, and if so, how many and when. The development of efficient methods of contraception is seen as contributing to enlarging persons' effective range of choice in an area of their lives that is of major importance to them. Also, of course, marriage and family life, where they are successful, are major areas of self-development. Consider Aristotle's discussion of the family as being necessary for man, because of his lack of self-sufficiency.

Against this kind of view, it is widely held that the rights to marry and have children are distinct from and much stronger than these rights. Certainly, exponents of the Thomistic theory of natural law have seen them as very important, basic rights. In those Papal encyclicals which are heavily influenced by Thomistic thought, there is great stress on the right to marry and have and rear a family, and this not as an aspect of a right to liberty. Indeed, Pope Leo XIII, whose encyclicals a century ago did so much to form, guide, and redirect Catholic thought about social justice, denied that we had a general right to liberty of the kind that liberals claim we possess. Yet he stressed the rights to marry and have a family as rights that no state could possess a moral right to invade. The argument appears to be that by virtue of our nature as animals, we possess a sexual inclination, the end of which is procreation. We have a right to attain our natural end, which comes from perfecting our natures, hence we have the right to marry and reproduce. To deny a person the right to marry, reproduce, rear his/her offspring, is to deny him/her scope to attain his/her natural end. Saint Thomas does not argue that we have an obligation to marry and have offspring, although his discussion of those who choose celibacy for religious reasons does suggest that there is a tension in his theory. In fact, he simply suggests that the natural course for those who do not have a religious vocation will be to marry and have offspring. The encyclicals of the past century relating to marriage and the family have all reaffirmed the general view of Pope Leo XIII, with the qualification that the most recent, relevant encyclical, that of Paul VI, *Humanae Vitae*, hints that there is a duty of responsible parenthood, in the sense of there being a duty not to produce offspring for which adequate care cannot be provided; this is in contrast with earlier encyclicals, including Pius XI's *Casti Connubii*, which indicated no duty to seek to contain the size of one's family, this being seen as God's perogative. In terms of the arguments relating to the rights to marry and reproduce, in spite of the dubious defense of the unreliable ovulation method of birth control, sexual abstinence would seem to be the only morally permissible method of birth control in terms of the teleological ethic underlying the views set out in the relevant encyclicals, Leo XIII's *Arcanum Divinae*, Pius XI's *Casti Connubii*, and Paul VI's *Humanae Vitae*. There are telling objections that may be urged against that ethic in terms of questioning its account of "good," and the teleology that is basic to it.

The right to private property is, in many ways, basic to the liberal society as we know it. Insistence on full respect for a right to private property would severely restrict the range of measures that would be possible for the sake of conservation, preservation, or redistribution of wealth to enable the destitute of the world to enjoy their basic human rights. On the other hand, many measures directed at checking pollution could be justified as measures preventing invasions of property rights. It is therefore important that it be determined whether there is a right to private property, and if so, what is its nature and extent. Many environmentalists deny that there is or can be such a human right; in particular, they deny the legitimacy of property rights in and to the environment, whatever that may mean.

The philosophy of the right to private property is one of the more muddled, ill-thought-out areas of moral and political philosophy. Even though, as a result of the work of political scientists and economists, philosophers are at last coming to see that private property is not one thing, land, but many (land, buildings, shares in corporations, rights of use, control, and disposal, rights to income, and the like), there is still a strong tendency to think about and argue for a right to private property on the basis of a simplistic view of the nature of private property. The most persistent argument is that of John Locke and so many other, that from the right to one's person, one's body, one's labor as an extension of one's person, the fruits of one's labor, and to that on which one labors; Locke and others qualify the right by reference to the requirement that one not take more than one can use and that one leave enough for others. Locke's formulation runs:

Though the earth and all inferior creatures be common to all men, yet every man has a 'property' in his own 'person'. This nobody has any right to but himself. The 'labour' of his body and the 'work' of his hands, we may say, are properly his. Whatsoever, then, he removes out of the state that Nature hath provided and left it in, he hath mixed his labour with it, and joined to it something that is his own, and thereby makes it his property. It being by him removed from the common state Nature placed it in, it hath by his labour something annexed to it that excludes the common right of other men. For this 'labour' being the unquestionable property of the labourer, no man but he can have a right to what that is once joined to, at least where there is enough, and as good left in common for others. (Locke, 1690, 26).

Although Locke discarded the labor theory of property almost as soon as he had completed his exposition of it, it has continued to influence thinking about a right to private property. There are difficulties with this argument at every step. It is not at all clear that one "owns" one's person. One's labor is not simply an extension of one's person; it has a social element in it, effective labor usually itself being a product of social training as well as of personal effort. The so-called products of labor, the fruit, corn, fish, land, are not such at all. Their being in the laborer's hands is the product of his labor. Nature creates the fruit, corn, fish, land. Labor may slightly change or modify land. The world is finite and its population is large, so the sufficiency condition could no longer be met, and, if adequate account had been taken by the original appropriators of property of the future rights of future generations, they could never have claimed or acquired permanent ownership rights to pass on to their descendants. Labor could not justify the right to bequeath property. We do not now believe that our "ownership" of our bodies entitles us to determine how they are to be disposed of after our death, so how could the mere products of our labor be more truly ours and be such to admit of the right of being bequeathed? Property ownership today is not based on the fruits of labor followed by free, voluntary, fair transfers, but on military conquest, theft, fraud, exploitation of slaves, intimidation of workers, and the like, and has little relationship with the distribution that would have resulted from laborers receiving the fruits of their labors. If the labor argument were to be taken seriously, we should need to start afresh in allocating property. Then the problems would arise in respect of the finitude of the earth, the social element in labor, the social value of what labor produces, the determination of the relative contributions of different laborers in the complex technological society, and the relevance of all these considerations to the many kinds of property and property rights. In brief, the theory is both false and one that it would be impracticable to seek to implement.

Locke did note another argument for private property, that from the right to life, and from what is necessary for the sustaining of life. This is a more promising, more serious approach to property rights. However, the right to life is not the only right that bears on property rights. I suggest therefore that what is a just and morally acceptable system of property ownership and distribution of property is to be determined by reference to what is dictated by concern for respect for basic human rights. If this approach to property rights is followed, we should deny that there is any basic human right to private property, and argue that property rights are to be determined in terms of what is most conducive to respect for other rights and other values. On that basis, there can be no moral bar to measures necessary for conservation, preservation, and the solving of other ecological problems on the basis that there is a natural human right to private property. At most, there will be constraints due to duties of justice to compensate fairly those who are disturbed in their present property rights by ecological reforms.

Resources: Depletion and Conservation

Much of the discussion of the case for preservation of species and wilderness was to the effect that preservation is morally desirable, even morally obligatory, when and to the extent that it is dictated by conservation of resources. The issue of conservation of resources arises also and is most commonly discussed in respect of resources of the kinds commonly referred to as "renewable" and "nonrenewable." In the discussion of the case for preservation it was assumed that it is not simply imprudent but immoral to waste resources, especially renewable resources. It now needs to be considered why, if so, the waste of resources is immoral, and if not all waste is immoral, what kinds of waste are such.

Much of the concern about wasteful uses of resources, especially but not only of scarce "nonrenewable resources," emanates not from a moral concern for conservation but from a moral condemnation of the luxurious, self-indulgent life that those in wealthier countries enjoy, and enjoy by "squandering" the resources of the world. This condemnation harks back to the moral ideal of moral asceticism. Linked with, but distinct from this, is the so-called puritan ethic, which condemns waste qua waste simply because it is waste and not because it leads to a luxurious life or to a neglect of duties that are owed to the destitute by virtue of their possession of the rights discussed above. A third view is that the present use of resources is unjust and wasteful in the sense that it is not related to serving man's best interests, or even man's basic needs. While in wealthier countries there is a great waste of resources, food and other resources, and also within the wealthy classes in the less affluent countries, the waste is often by those parasitic on the exploitation of the destitute, those suffering poverty, malnutrition, even starvation, and a lack of the amenities essential for a life with dignity and enjoyment of basic human rights. The waste is such as to achieve no worthwhile moral good to justify the gross lack of respect for the moral rights of hundreds of millions of persons it involves. If the rights of these persons are to be respected without the elimination of the present wasteful use of resources, this result could be achieved only by vastly increased production of food and other goods, that is, by using more and even more of the earth's resources and thereby possibly risking the exhaustion of various nonrenewable resources, damaging or reducing other resources such as wilderness and species, and increasing pollution, including the very dangerous pollution due to nuclear fuels and nuclear wastes, which in turn may render renewable resources such as land, sea, and air nonrenewable. Simply to increase food production without improving the method of distribution would provide no lasting solution. The wasters would simply waste more, with relatively little trickling down to those who most need it. Only massive increases in production would cope with the increased waste, while providing the needy with the food and other resources to which they have a moral right. To many, it seems evidently morally preferable that conspicuous waste be eliminated, rather than that these evils be tolerated. Others argue that waste is morally quite unjustified, quite apart from the ecological effects of allowing it to continue unchecked, that it is immoral because it rests on injustice and a failure to respect the rights of the destitute.

Much of the alarmist writing concerning the possibility of exhausting key resources and thereby endangering human survival is basically directed at gaining acceptance of an ascetic morality, and this for its own sake, and not for ecological reasons, and not for reasons related to respect for human rights. Telling moral arguments, which need not be rehearsed here, are available against asceticism as a moral point of view. They include arguments that relate to the value of human self-development and the exercise of personal autonomy. Nonetheless, while luxury and waste are not necessary for effective self-development, the context in which they prevail is that in which those who enjoy them are most likely to be able to be self-developing and to exercise their moral rights effectively. Similarly, the condemnation of waste qua harmless waste is easy to expose as morally groundless. To use solar energy "wastefully" to overheat or overcool a building, to power cars used for purely recreational travel, or the like, is in no way morally wrong, even though the uses to which this solar energy is put may offend the ascetic. What makes the waste of today as immoral as it is is the context in which it occurs, namely, one in which many are in desperate need, and in which waste occurs when the resources could and ought to be used to provide help to which the hundreds of millions of persons concerned have moral rights. That it is a wasting of scarce resources for which future generations may have a real need makes it all the more indefensible. The moral concern about waste is therefore a wider, more embracing concern than simply that about the moral desirability of conservation of resources for use by future generations.

The moral concern about conservation relates to nonrenewable resources, and the danger of rendering renewable resources, the land, air, oceans, animal and plant species, and wilderness, nonrenewable. Discussions here center around the increasing—according to many, the exponential—growth in the rate of use of resources, including the "nonrenewable" resources, attempts being made by some to predict when the reserves of fossil fuels, various metals, and other raw materials, will be exhausted. Further, it is argued that even if the exponential increase in the rate of use of many of the nonrenewable resources were to end now, and the use stabilized at the present levels, important resources of which we and future generations might be expected to have a need will, in the not too distant future, be exhausted. To raise the standard of living of those in the underdeveloped countries to one nearer those enjoyed in the countries of the Western world would, in this view, be seriously to aggravate the problem with major harmful ecological implications. It is claimed that each person in the United States uses an average of eighty times the energy used by a person in India, and about twenty times the resources used by persons living in underdeveloped countries. Given the seemingly insatiable nature of human demands, such that what today is a luxury is tomorrow a "necessity," and that all aspire to attain a higher and still higher standard of living, and hence to use more and even more nonrenewable resources, the possibility of using up certain key resources and thereby creating a crisis in living in the high-consumption countries in particular, but also in the world in general, is presented as being a very real one. That an exponential growth in pollution is claimed to go hand in hand with this exponential growth in the use of scarce resources is claimed to make the crisis in living that faces mankind an even graver one.

Many forecasters of a resources crisis see the crisis as one of survival for the human race, avoidance of this being dictated either by species prudence or by morality, or by both. Others stress the massive suffering, starvation, degradation, that would come from a breakdown of our economic and social order, and stress the duty to prevent this. The latter duty is exactly the same kind of duty as is the duty toward the destitute today, the difference being simply that the one is toward persons now living, the other toward those yet to be born. If there is a duty to conserve and to avoid wasting nonrenewable resources for the sake of future generations, there is equally a duty to make these resources available today to those who need them.

Most discussions of the case for conservation see the issue as one of the needs and satisfactions of the present generations against the possible needs and rights of the generations of the future. That is not how the problem arises, if at all, in the concrete. Today the rights of the majority of the world's population are not respected.

Many demands for conservation are demands that we conserve resources without doing anything about this injustice. They are demands that we cut back the rate of growth of the use of resources, or, more drastically, that we peg total resource use, even that we reduce it, in respect of all, some, or few very scarce resources, where the assumption is that this can be achieved only by allowing the present injustices to continue, since to bring about enjoyment of their rights for all mankind would be vastly to increase the rate of use of scarce nonrenewable resources. Insofar as that is what is being demanded when conservation of resources is demanded, there can be no moral case for it. That is to prefer possible needs and possible rights of possible future generations to the actual needs and actual rights of living persons. In such a moral calculus, the actual moral needs and rights of actual persons must outweigh the possible needs and rights of an unknown, unknowable number of future persons. We do know what are the needs and rights of the living. For reasons of war, disease, natural disaster, climatic changes, there may be few or no human beings at any given future date. Further, we do not know what raw materials those who do come into being will need as resources, since what is a resource depends on socio-economic-technological factors. Hence we may conserve raw materials that could be used with advantage by the destitute today but come not to be seen to be or to be used as resources by the future generations. If solar energy were to be harnessed, many, possibly even all, existing energy resources might cease to be such. It is the rights that future generations will possess that morally are of importance. What respect for them dictates must be a matter of great uncertainty, given our lack of knowledge as to their numbers, the technological advances that will occur in the future, the lifestyles that will be favored, and so on.

The uncertainty about the needs of future generations tells against conservation at the expense of the satisfaction of the needs of present generations. Our uncertainty as to whether there really is a resources crisis, and also about the contribution conservation of resources can make toward preventing it, if one really threatens us, also tell against conservation at the expense of present generations. These considerations are related. Claims that a resources crisis is imminent rest heavily on stress on the earth's finitude. The rejection of these claims usually goes with a stress on the immensity of the earth, on how little of it has been explored and exploited in respect of resources—only the superficial land surface, and little of it, not the oceans, not Antarctica, not land masses at any great depths. Also stressed are the facts that the bounds to our reserves are economic, technological, and, in the case of energy, moral, rather than physical.

The whole point of conservation as a moral exercise depends on technology making advances, on man, through technology, finding alternatives to the scarce, nonrenewable resources. Conservation cannot prevent the exhaustion of resources and the crisis that must follow, if a crisis must follow the exhaustion of various of the nonrenewable resources. At best, it can simply put off the evil day. If, meanwhile, population increases, then conservation, by putting off the crisis, may bring about vastly greater misery, and a much greater catastrophe. Obviously those who favor conservation do not believe that they may be setting things up for a greater calamity; they do not believe that conservation will simply postpone the evil day. Like their opponents, they too believe that man will find alternatives to the presently key, scarce nonrenewable resources. They believe that we need more time, that it is prudent to give ourselves more time, in which to allow ourselves to find these alternatives. They too have faith in there being the necessary scientific and technological discoveries. That is a reasonable, morally proper stand to take, given the seriousness of the consequences, should a resources crisis eventuate. However, it would not be a morally proper course to take if the conservation were to be at the expense of satisfaction of the needs and rights of those now living. On the facts available, it would seem that if resources were justly distributed today, it would be possible to secure the enjoyment of their rights by all and also to curtail growth in the use of nonrenewable resources.

This would not mean opting for no growth, as growth using renewable resources and resources which are in such abundance and which admit of being recycled would be possible, although concern for conservation of energy resources would be relevant here. Much that is valuable and worthwhile in present-day life, and in our civilization and culture, is compatible with a curtailed use of scarce nonrenewable resources, with the kinds of economies in the use of resources that were practiced by the civilian population in Britain during World War II. Further, given that above a certain minimum standard it is not the absolute standard of living but relative standards and relativities that bear greatly on personal satisfactions, a less luxurious style of life if freely adopted need not be a less satisfying one.

If this is so, what then is the strength of the moral case for opting for conservation and an austere mode of life? If voluntarily entered into by individuals, the cost to each person could be small, while the benefit to future generations could be very great. This might suggest that there is a clear-cut, overwhelming moral case for us as individuals and as a society, opting for conservation. However, because of our lack of knowledge about the future, and because a policy of conservation would mean certain losses of goods, including a higher level of unemployment, against simply possible goods for persons in the future, the case is not as clear as it might seem to be. The difficulty our lack of knowledge creates becomes more evident if we consider the next question that must be answered, if a policy of conservation is adopted, namely, that concerning how much conservation we should engage in. To engage in too little would be to forgo goods and achieve no gains. To engage in too much conservation is to forgo goods unnecessarily, with no gain to future generations coming from our overconserving. Yet we cannot know what is the right amount of conservation to engage in, except that it will vary from resource to resource. At this point it might be suggested that there is an easy answer, that we should avoid all wasteful uses of nonrenewable resources. However, the concept of a wasteful use is not a clear, well-defined one. It is possible to view as wasteful all use of resources in excess of a very austere mode of existence—for example, all use of petroleum for purposes of pleasure, all use of energy to heat buildings to more than 16°C (60°F) or to cool them below 30°C (86°F). Our ignorance as to the needs of future generations must make any decision here arbitrary.

Given these considerations, I suggest that no case for conservation in general in the abstract is possible, that it is necessary to approach conservation resource by resource. The case for conservation is strongest where a resource is a key, nonrenewable one presently in limited supply, such that on the basis of known reserves and our knowledge of the thoroughness of our search for further reserves, it could be exhausted in the foreseeable future, with consequent hurt to the social and economic system and to those living in it. This would seem to be the case in regard to petroleum today. However, this approach gives no guide to what savings, what reductions in use, should be aimed at. With other resources, for example, coal, the time will come when conservation is dictated by concern not that our energy resources will be depleted but that we will become dependent on nuclear fuels with the risks of nuclear accidents, pollution from nuclear wastes, and the like. Man needs to allow himself time to develop the less dangerous, natural energy sources. It is not clear that there is presently a case for conservation of coal, although it is obviously imprudent for man to use such a valuable resource for trivial ends. In brief, the case for and against conservation needs to be investigated in respect of each scarce nonrenewable resource.

By contrast, based on current political realities, a strong moral case can be made against conservation in the abstract. There is no real hope of bringing about a just redistribution of resources in the world in the immediately foreseeable future. To ensure that the unjustly treated majority of persons in the world today will enjoy their rights, we must greatly increase resource use in the Third World. If that occurs, conservation could then be achieved only by massive cutbacks in resource use in the affluent countries of kinds that are socially, economically, and politically such as to be practically impossible to achieve. Given these facts, the demand for conservation will in effect become what it already is in the writings of some of its exponents, a demand that resource use be pegged as near present levels as possible, with roughly the present distribution.

The case for conservation in the sense of keeping renewable resources renewable is very different. We clearly have a duty to future generations, near and distant future generations, to keep renewable resources renewable, and hence available. With most such resources, this simply restricts our misuses of them. For example, it restricts our overusing or misusing land. However, it also bears on what we do to the environment by way of enduring pollution, as pollution can make oceans, the atmosphere, and land areas no longer resources. And it and other causes can lead to useful species of plants and animals becoming extinct.

The question of our duties as individual persons is very commonly raised, it being suggested that it is immoral not to conserve, although not all possible conservation of resources is seen to be morally obligatory. Most of those who interest themselves in the morality of conservation live in high-resource-use countries. Most such persons are themselves high users of resources, including scarce energy resources. Those of us who live in high-resource-use communities, by our use of planes, cars, boats, and so forth contribute to the depletion of petroleum; by our use of household appliances and facilities, TV, refrigerators, freezers, washing machines, dishwashers, and the multitude of other gadgets including heating and cooling systems in our homes and offices, and equally by our demand for goods manufactured only with the use of vast amounts of energy, we contribute to the depletion of other energy sources, coal, nuclear fuels, natural gas, and bring nearer the day when there will be real dependence on nuclear fuels, if natural energy sources are not effectively harnessed. These facts raise the question of whether we have the duty to reduce our own personal use of energy resources, as by using inconvenient, unreliable, time-consuming public transport, and, more generally, by living in an energy-frugal way. If we were to accept and apply the once very popular utilitarian argument expressed in the question "Suppose everyone did the same?," it would seem that we ought so to order our lives, no matter what the actual consequences of our changing our mode of life. Equally, if we applied the universalization principle, it would seem that we could not possibly act as do most ecological moralists. In spite of such considerations, I suggest that it is only where our conduct is genuinely influential that we ought to restrict our use of scarce resources, since the losses to us in changing our lifestyle outweigh the benefits achieved by adopting a self-denying lifestyle. Given the kind of society and social life, urban development, and the like that have developed in high-resource-use countries, we should stunt our personal development and lives as individuals, if, in such societies, we sought to live resource-frugal lives. On the other hand, if resource frugality as individuals contributed to reducing the evils of the present distribution and use of resources throughout the world, there would be a case for it, not so much from conservation for future generations but from justice to present generations. However, there are good reasons for believing that it is by political action, new economic policies, removal of tariffs and quotas, opening up of markets, fair trade and competition, and freer immigration as well as well-planned technological and other aid, with the sacrifices in living standards that the former would bring to persons in affluent countries, that those injustices are best to be lessened.

Science and Technology

A useful definition of technology is that given by E. G. Mesthene: "We can define any new (nontrivial) technological change as one which (i) makes possible a new way of inducing a physical change; or (ii) creates a wholly new *physical* possibility that simply did not exist before." And: "It brings about or inhibits changes in physical nature including changes in patterns of physical objects or pressures" (Mesthene, 1968, 135).

Science and technology, while not resources in the usual sense, may usefully be discussed in the context of resources, both because what are resources usually become such because of scientific and technological discoveries and because science and technology discover or create many new resources. Consider here the development in the use of aluminum as a resource, and the invention and use of plastics, synthetic fibers, detergents. Technology, and the science on which it rests, thereby enable us to replace, often with improved alternatives, various formerly essential, key renewable and nonrenewable resources with other renewable and nonenewable resources. Science and technology do much more than this. They confer on man great powers that he otherwise would not possess. They thereby confront him with new, important moral decisions about whether to use these new powers and, if so, how and when, and in what ways. Besides conferring on us powers in respect of the exploitation and use of resources, creating new resources, new products, new tools, technology has lessened our dependence on human labor and thereby provided greater scope for that leisure that Aristotle so rightly believed to be essential for attaining excellence as a man. In Aristotle's state, this leisure was achieved not by technology but by slavery. Science and technology have provided knowledge that bears on most aspects of social life and the development of society. They have provided knowledge that has increased life expectancy and lowered death rates; they have given us increased control over birth rates and hence over population sizes; and in innumerable ways they have greatly contributed to increasing food production. Indeed, most that we value in the world today, including our written language and all that that makes possible, are products of technology. A pretechnological society would be one without the controlled use of fire, the wheel, all tools, machinery, medicine, written language. It would be very like Hobbes's state of nature, except that in the latter, the only human society is that of the family.

These facts are of great importance given the growing distrust of technology and the ventilation by many ecological moralists of the view that we should be better off with much less technology. Technology and science cannot be put aside, rejected, or checked in their development without serious loss to man's well-being and future good. It is romantic nonsense to think that man could turn his back on technological inventions. Equally, it is arbitrary to suggest that although technology up to now has been beneficial, we should reject all further technological advances. Nonetheless, there are reasoned expressions of moral concern and distrust of technology that merit mention here.

Reasoned concerns about technology include the claims that technology, which, because of the additional powers it confers on us, ought to enlarge our freedom, can in fact lessen our freedom; technology brings us powers that we lack the judgment to exercise wisely and well, and which are extremely dangerous if possessed by those who lack the relevant wisdom and judgment. Concern is also expressed that the more complex and powerful the technology, the greater is its effect on nature, and the more likely that its effects will be harmful and even dangerous. It is also contended that technology is not morally neutral. A further concern is that the more advanced our technology, the more exposed we are to being dominated by a technological elite.

At first sight at least, technology would seem very evidently to have enlarged our positive freedon and our effective range of choice and thereby contributed greatly to making us more the masters of our destiny. Technology has made available a range of choice in living standards. It has extended our choice of materials, resources, products. It has opened up new, more efficient methods of food production. It has developed transportation, which allows us freedom of movement, which in turn opens up many more avenues of choice and freedom. The development of communications has made the world really one community, such that, if man had the will go do so, he could move food and other essential supplies quickly to those areas affected by famines and natural disasters. Developments in respect of the media have made knowledge and culture accessible to hundreds of millions of persons. And medical technology, besides reducing death rates and giving control over conception, has contributed to lessening many of the incapacities that were occasioned by disease and injury. Given all this, it may seem surprising to find it claimed that technology restricts our freedom—according to some, that it may even enslave us. Consider here J. Ellul's observation that "in the modern world, the most dangerous form of determinism is the technological phenomenon" (Ellul, 1964, xxxiii; see also Watson, 1971).

The thought is that technological processes develop their own momentum, they have their own motivations and goals, they gain a cumulative force from the coming together of many decisions, many individual uses of technology, creating a social and economic force that may encompass and even overpower individuals and individual decisions. Clearly all this is true. Further, obviously, technology can alter the structure of society and thereby limit choice. We can indeed become obliged to use technology we should prefer not to use. Consider here the effects of mass production methods in industry on the would-be craftsman, the necessity of the ordinary person in modern society to use cars, jet planes, computers, to possess and use phones, and innumerable other appliances and gadgets, even though many would wish to be free of their "dependence" on the car, the phone, the plane. The impact of technology on freedom is not all one way. For some individuals, there may well be a total loss of freedom, even of positive freedom to be effectively self-determining. However, on balance it would seem that there is a clear gain in the effective range of important choices for most of mankind, although there clearly are losses of specific freedoms.

The view that technology brings us powers that it is dangerous for man to possess and to exercise is one canvassed by many, for example by D. Lyons in "Are Luddites Confused?" (see Lyons, 1979). There Lyons expresses concern at the dangers inherent in further empowering fallible humanity. The burden of his argument relates to the possibility of the misuse of technical powers, that such a possibility is likely to be realized with serious damage to world ecosystems and to their adjustment mechanisms. He argues that the technical know-how develops too fast, to be wisely and responsibly exercised by man with his limited wisdom. Lyons raises important matters of concern. Man has very often been unaware of or underestimated dangers inherent in the use of technology. He has seriously misused many technological inventions. It is true that men unfit to exercise power come to possess great power, that evil men such as Hitler, Stalin, and rather too many leaders of that ilk have had access to powers the misuse of which it is frightening to contemplate, and that democracies have had senile leaders, psychologically sick leaders, and corrupt leaders who have been ill equipped to make decisions concerning the use of nuclear weapons, germ warfare, and the like. Man is fallible. Technology has unleashed great powers. These powers have and will in the future fall into the hands of reckless, irresponsible persons, and not simply political leaders. They are such that even responsible persons may misuse them due to ignorance, misjudgment, careless-ness, thoughtlessness. All this is true. Yet in spite of all the risks of harm and damage to man and his environment that the powers technology has conferred on man this century bring, it is noteworthy that in the thirty-seven years since World War II the misuse of these powers has been severely limited. It is true that man does, by and large, learn more often than he ought by trial and error, but it is also true that mankind collectively acts much more wisely than one would expect by considering men individually. Mankind collectively has exhibited much more restraint, judgment, even wisdom about major matters in his use of technology than might reasonably have been expected from a consideration of his capacity for misuse of this technology.

Part of Lyons's concern relates to the effect of technology in bringing about great, harmful ecological changes. His thesis is that the kind of great, random changes that come with the use of technol-ogy in affecting ecosystems are much more likely to be damaging to the ecosystem; he echoes Commoner's theme about the general harmfulness of arbitrary interventions. Lyons's concern is well di-rected, but it needs to be balanced against the fact that this kind of danger is more generally appreciated, that man can and does take increasing care in his use of such technologies, and also that nature is much more resilient than it is now fashionable to acknowledge. Consider here the resilience of nature and the stability of the ecosys-tems of factory farms, feedlot farming, irrigation areas, areas with new varieties of cereals, the ecosystems of pine forests which have replaced eucalyptus forests, and the so-called monoculture ecosys-tems.

The fourth concern relates to the alleged lack of moral neutrality of technology. Ellul, Watson, and Lyons all touch on this issue. I suggest that, in the abstract, Watson is right in claiming that technology is neutral. However, as Ellul notes, there is a tendency to think that we must use any new technological powers; and this may lead to unnecessary, undesirable, morally unjustified uses of technology. Consider here the unnecessary, undesirable uses to which modern life-support technology and other techniques of modern medicine to keep "alive" worthless organic life, have been put. Genetic engineer-ing may well similarly be misused. The uses of the former technolo-gies are increasingly coming under critical scrutiny, such that even the most conservative moralists are coming to agree that it may morally be better in certain cases not to make use of the powers such technology makes to be available. Technology does make great wick-edness, great evil, possible. It also makes to be possible great goods. It provides power. How mankind uses that power is up to mankind. Man is morally fully accountable for how this power is used. He cannot plead lack of responsibility on the grounds that technology itself is not morally neutral.

The fifth concern about technology is the very important one relating to the danger of being dominated by a totalitarian technlogi-cal elite. Of this R. A. Watson writes: "This is close to saying what people as diverse as Tom Hayden, George Wald, Eldridge Cleaver, Benjamin Spock, and Noam Chomsky have been contending on the grounds of similar humanitarian principles: The organization of insti-tutions to use modern technology to serve the needs of all mankind will require something like a revolution. The fear of such a reorgani-zaton is that centralized control of technology either will lead to totalitarianism or be achieved by totalitarian means. A totalitarian future for man is in fact anticipated by Ellul as inevitable given the present course of technology" (Watson, 1971, 224). Those who make this kind of point are right about the dangers that come with compu-ter technology and genetic engineering, as well as the kind of mass production supported by computer technology. Those who control the computers and who determine the decision making used in the operation of modern technologies will have very great power. How-ever, while this could result in the emergence of a technological elite of a kind that is morally and politically unacceptable, there need not develop a stable, unchanging, nonresponsible elite of the Platonic philosopher-king kind. Rather, it is vital that we work to achieve a well-run society in which there is a technological managerial class that is subject to control by the democratically elected representatives of the people, and answerable to the community. Nonetheless, the problems posed by specialists with powers for which they need not be accountable to the community, and of a totalitarianism of bureau-crats, are real ones that need now to be faced and met. There is no reason to believe that the liberal democratic state cannot devise the controls that will be necessary.

In brief: Our social and personal duties in respect of science and technology are determined by an informed application of the moral values set out and defended in this work. There are new practical problems to be faced, but the problems do not call for a new ethic. Rather, they call for an informed, vigilant application of the ethic that centers around respect for persons, justice, honesty, and promotion of good and respect for human rights.

Pollution

Pollution raises important moral issues of respect for persons, justice, honesty, and prevention of evil, involving as it does at the least a nuisance and commonly a threat to health and life of those now living and/or of those of future generations of persons, as well as harm to animal and plant life, wilderness, natural phenomena, and to artifacts of great value. Pollution is not a new phenomenon, and the moral issues that it raises are familar ones capable of being dealt with within traditional ethics. Ever since man has lived in society, there has been man-made pollution, including dangerous forms of pollution. Garbage and sewage have long constituted forms, often dangerous forms, of pollution, exposing societies to serious diseases as well as the nuisance of pests and the pollution of smells. Atmospheric pollution from coal is centuries old. What is new, and what rightly is a matter of great moral concern, is the range of kinds and forms of pollution, its often insidious, dangerous character, its possible long-term effects, and more generally, that pollution may occur at places distantly removed from the source of the pollution in space and time. There is air pollution of many kinds, from many sources, from fossil fuels, nuclear energy, nuclear weapons, pesticides, industrial air effluvium; there is water pollution from sewage, pesticides, industrial poisons, fertilizers, nuclear fuels, heat. There is pollution that affects the land itself. More generally, serious, dangerous pollutants today include a vast range of states, conditions, substances, such that unless we are unusually fortunate the water we drink, the air we breathe, the food we eat, the environment in which we live, will at least sometimes be polluted, even seriously polluted, by pollutants that endanger our health and lives. As noted in Part I, ecologists stress the dangers that arise for man from many forms of pollution from the interconnectedness of things in nature, the build-up into dangerous concentrations of poisons in food chains and the like.

Many moral issues are raised by pollution. All are made to be difficult because of the complex nature of the causes, occurrences, and effects of pollution. One aspect of the moral issue relates to the duty not to cause or unnecessarily to cause pollution, the causing of pollution usually involving a breach of the duties of respect for persons, justice, honesty, and prevention of evil. Another aspect relates to the morally best, morally most acceptable methods of controlling pollution. Questions of justice as well as of prevention of harm arise here. A third aspect relates to the allocation of responsibility for causing pollution, given that much pollution results from the actions and demands of many persons. A fourth aspect relates to the proper response by way of punishment and reparation for pollution. Yet another moral aspect of the problem of pollution concerns whether and when the harmfulness of the pollution makes an activity no longer morally acceptable. This is an issue that arises in respect of nuclear pollution.

In the abstract, it is clear that there is a duty not to cause pollution unnecessarily, since to cause unnecessary pollution is to cause harm needlessly and to show lack of respect for persons. Often it is also thereby to act dishonestly and unjustly in that the costs caused by pollution are "externalized" by the polluter and made to be borne by others who ought in no way to be called on to bear them. In the concrete, things are much less clear, both because what counts as "unnecessarily" causing pollution is open to debate, and because much serious pollution is the result of the cumulative effects of small, seemingly harmless amounts of pollutants. It is impossible to live in society today without significantly contributing to pollution of various kinds. We may, do, have the moral right to pollute, if this is necessary for our survival and well-being, and if the pollution does not cause serious harm to others. More typically, the situation is more complex than that. The individual's contribution to pollution may be slight, but equally the claim that we *must* engage in the activity that causes the pollution is often questionable. It usually is possible to reduce the amount of pollution to which we are contributory causes, as by using public transport and not cars. With other polluting activities, such as the use of pesticides in food production, the individual may knowingly cause serious pollution, yet the price to be paid in terms of loss of food if it is stopped may be too high to be morally acceptable. That so much pollution in isolation is of no moral significance, but in conjunction with that caused by others, as with leaded gasoline, may be very seriously harmful to some persons, makes for difficulties in making a judgment as to when the pollution is "unnecessary" and about the allocation of responsibility for it. That some pollution may be harmful to future generations, and not only or even not seriously to living persons complicates the moral calculus. The limitations of our knowledge as to which pollutants are in the atmosphere, the seas, rivers, about their effects, even whether they really are pollutants, and if so, dangerous ones, make for further problems in resolving the moral problems of pollution.

Many environmentalists approach the problem of pollution with a sense of moral outrage and see a political response as the appropriate one. They see that many polluters are harming us, poisoning us, endangering our health and even our lives, and believe that no one can have a moral right so to act. Hence the polluter must be stopped by being rendered a criminal and punished as a criminal. Such an approach rests on an oversimplification of the issues. While both punishment and the extraction of reparation from polluters are often morally appropriate responses, they are not always so. Equally, they are not always the most effective responses, if our object and our moral goal is to reduce or prevent the pollution. The polluter may not be aware, and may not even be able to come to know, that he is contributing to the production of pollution. Or he may be obliged to pollute in order to live and to produce, as in the case of the farmer who must use pesticides against weeds and pests to secure his crops. The manufacturer whose factory pollutes the atmosphere and the river may be employing a large number of people who will lose their jobs if he is forced into uneconomic production costs by adopting antipollution measures. The provision of employment for large numbers of persons in an impoverished area may depend on the acceptance of energy plants that cause significant pollution. And so on. Thus, while in the abstract the most morally appropriate response to pollution may seem to be that of punishing the polluter, in the concrete many other responses are seen to be morally preferable. If the pollution is a lesser evil, and one to which those affected consent, it is morally more appropriate for the state to seek to help the polluter reduce the pollution, contributing to the costs if need be. The nature of the appropriate political responses will be discussed in Part III. Here it simply needs to be noted that a concern for what would seem to be dictated by strict justice is not always the most effective or morally the best-justified approach to the reduction of pollution.

In any case, there are problems in determining what justice dictates, in particular, in justly allocating responsibility for pollution, as much pollution, as from cars, detergents, pesticides, fertilizers, and the production of electricity, results from the coming together of the effects of the actions and demands of many persons, where, if each individual were the only person concerned, there would be no pollution. The many problems in justly allocating responsibility set limits to what can properly be imposed as just, retributive punishment, and what exacted as just reparation. On the other hand, where the pollution is seriously harmful to health, well-being, or to life, and significantly interferes with the enjoyment by persons of their human rights, and responsibility for the pollution can firmly and accurately be laid at the door of a specific person or group of persons, including such legal persons as national and multinational corporations and their officials, criminal action, retributive punishment, and the extraction of just reparation are morally appropriate, desirable, and morally necessary. Justice demands such punishment and reparation, and such protection of human rights. These measures are also dictated by other moral principles and concerns. The production of unnecessary, avoidable, harmful pollution constitutes serious lack of respect for persons. Further, such polluters are dishonest, imposing as they do costs incurred by their activities on others who are in no way party to them. And they often in the process damage valuable artifacts and natural phenomena. What makes the resort to inflicting retributive punishment so attractive to many is the fact that many polluters, including states and corporations, show a callous lack of respect for human life, human persons, and human rights, permitting pollution they could easily avoid or prevent. Multinational corporations and their officials commonly, knowingly, for the sake of greater profits, permit high, dangerous, unnecessary pollution levels that seriously endanger health, well-being, and enjoyment of human rights, varying the levels of pollution they cause and permit according to what they can best get away with economically, given the laws and the

enforcement of those laws, and their penalties, in the various countries in which they operate. Many who so produce dangerous,
damaging pollution, and this quite unnecessarily and simply to
increase already great profits, the officials and the corporations are
and ought to be treated for what they are, mutilators, violators,
deformers, murderers of persons.

The extracting of adequate reparations for harm done, where it is
appropriate that reparations be sought, encounters problems not
simply from determining degrees of responsibility, but also in assessing the costs of the pollution to individuals and the community. How
is the responsibility for pollution in Rio de Janeiro to be allocated
between the officials in Rio de Janeiro and the officials in the head
office of the corporation in New York, London, Geneva? Given the
vast variety of kinds of damage done by pollution to property, to
persons, to their health, to their enjoyment of life, and to the length
of their lives, computations of damages in terms of dollars is extremely difficult, and such as must be unsatisfactory and appear to be
unjust to the injured parties. This is a problem that is not peculiar to
harm due to pollution. Claims for compensation for injuries in
industrial and motor accidents raise similar problems.

As noted earlier, we have important duties in respect of future
generations concerning pollution. In discussing conservation, it was
noted that our uncertainty about the needs of future generations
seriously weakened the case for conservation, especially if it was at
the expense of the worst-off persons living today. In respect of much
pollution and the harm that it may cause to future generations, there
is generally less room for uncertainty. Pollution by nuclear radiation,
pollution from nuclear wastes getting into food chains, pollution that
leads to adverse climatic changes, pollutants that take a long time to
break down and persist in food chains from one generation to
another, may all seriously harm future generations. The duty not to
harm is a strong and determinate duty. It is generally acknowledged
by moralists other than utilitarians to be a stronger duty than that to
render help. It provides good moral reasons for not causing and for
not permitting many forms of long-enduring pollution and pollution
with long-enduring effects.

As with the conservation of resources, there are real problems in
determining our personal duties, because in many contexts our
polluting or not polluting will have negligible effects. We obviously
have a duty to abstain from pollution that harms another. However,
most of us are rarely in a position in which by our own actions alone
we are likely to cause seriously harmful pollution. More typically we
contribute to pollution in a multitude of small, indirect ways, as with
our use of energy, cars, jets, detergents, by our demands for newspapers, journals, books, attractively packaged goods, goods produced
by processes that cause dangerous pollutants to be released, or to be
used, as with foods produced using pesticides. Because it is the

collective action that is damaging, and because it is clear that if we as individuals cease to cause this pollution either by our actions or our demands for goods and services while others do not do so, no significant gain will result, while there will be some impoverishment of the individual's own life, it is hard to see that morality demands such individual self-denial other than where the pollution caused by the individual would itself alone be a significant hazard or contribution to a pollution hazard. It is through collective action, collective action best achieved through political action and support, even demands for such collective action, that a moral solution to the evils of pollution must be found and our moral responsibilities met.

Population

Most writers who warn of "an ecological crisis" see overpopulation as a major factor, as itself a part of the crisis ("standing room only") and as contributing to the other aspects of the crisis by way of speeding up the depletion of resources, destroying wilderness and with it rare species of plants and animals, and seriously worsening pollution of many kinds. Hence it is argued that there is an urgent moral obligation to contain population growth and to realize Zero Population Growth, some urging this as an immediate goal (not noticing or notwithstanding the disruptive concertina effects this would have on the population age groups, with varying demands for schools, employment, facilities for the aged, and the like), others as a more remote goal to be achieving without such uneconomic, varying age patterns in society. Others, including R. A. Watson and P. M. Smith, argue that the world is already grossly overpopulated; Watson and Smith in "The Limit: 500 Million" urge as a goal a world population of as little as only 500 million (Watson and Smith, 1970, 25 27). By contrast, the United Nations Fund for Population Activities (UNFPA) in its statement "The State of World Population 1981," noted the declining fertility in the world and predicted that the world population could stabilize in about the year 2110 at 10.5 billion, although it noted that it could be as high as 14.2 or as low as 8 billion. In its "Condensed Background Notes For Journalists," drawing on a recent study by the University of Wageningen based on the new FAO/UNESCO soil maps of the world, it suggests that world food output could be significantly increased so as to provide adequately for a population even larger than the highest of the three figures cited here.

Many questions arise therefore in respect of population. Besides the fact-value question of whether there is or is not a danger of overpopulation, there are many purely moral questions, as well as other important factual questions to be resolved. They include such questions as the nature of the individual's rights and duties in respect of having offspring, what are the morally permissible methods of restricting the number of offspring a person has, if there is a duty to restrict their number, what methods of checking birth rates the state morally may impose on individuals, if there is a danger of overpopulation, and what, among the morally permissible methods for individuals and states to adopt, are most likely to be successful in restricting population growth.

In Part I it was indicated that it was proposed in this work to use *overpopulation* to refer to population sizes that exceed a country's or the world's capacity to feed adequately and to make it possible for all to enjoy their basic human rights. A population that meets such a requirement need not be the optimum population. A smaller population may permit a richer, more widely shared culture as well as a higher standard of material living. However, because it is overpopulation that has been the major object of concern of ecological moralists, and because many of the same basic moral issues would be encountered by those who seek the realization of an optimum population size, the discussion here will focus on the moral issues relating to overpopulation. However, it is important to note that even if there is no danger of overpopulation in the foreseeable future, there may well be a moral and political case for checking population growth and aiming at an optimum population size.

Is the world overpopulated or in danger of becoming overpopulated? While it is clear that many countries for many different reasons—poverty of resources, lack of development of resources, mismanagement of their affairs, unfair world trading practices and arrangements, and the like—are overpopulated in the sense of being unable to feed adequately or generally to make possible for their members the enjoyment of their rights, the world as a whole is not overpopulated. It is possible to approach this question either by considering the number of starving and malnourished persons there are in the world today or by looking at the world's food production and potential and its capacity to meet other basic needs. In following the first approach one encounters widely discrepant figures. In writings throughout the last decade the figures that were most widely cited were that in a world population of 3.6 billion in 1970, 0.5 billion were starving while 1.5 billion were suffering from malnutrition. In sharp contrast, the UNFPA, in the "Condensed Background Notes For Journalists," produced in association with its "The State of World Population 1981," states: "The number of malnourished people in the world today is estimated at 450 million or about 10% of the world's population. 200 million are children under 5 and 10 million of these are severely malnourished." In the alternative approach, that of considering the world's food production and potential, discrepant, although not such widely discrepant, figures are also to be encountered. However, it is widely accepted that if the food actually produced were distributed and used to the best advantage of all persons in the world, and if the world did not curtail as it now does but increased food production to its maximum capacity, there would be more than ample food for all. Contrary to Malthusian predictions, food and other resources have more than kept pace with population. Further, were mankind to switch to a more predominantly vegetarian diet, but not to exclude meat entirely as a food, much waste of food due to losses in the conversion of plant protein to meat and in less productive use of land for grazing rather than for cereal crops could be avoided, and world food resources be vastly increased. There would, of course, be a price paid in increased pollution.

Is the world in danger of becoming overpopulated? Clearly there is and always will be a possibility of overpopulation in the sense that human beings have the capacity to reproduce at rates that could cause rapid increases in world population of a kind that could outstrip the world's capacity to produce food and other necessities. However, here we are concerned not with mere possibilities and capacities but with probabilities. Moral decisions of the greatest importance hang on what are the probabilities, not possibilities, in this matter. The only reasonable answer to the question is that we do not know. Those who so confidently predict a "population crisis" stress the rapid growth of the world's population, and often, as noted in Part I, predict simply by projecting the relatively short-term recent exponential growth rates into the future. This is totally unsatisfactory. To predict the future populations of specific countries and of the world at large accurately and scientifically is an extremely difficult enterprise. Experts in the field of demography rightly are very cautious in their forecasts. They are also notably fallible. The comparatively recent rapid increase in the size of the world's population is related to advances in medicine and a fall in death rates. In the wealthier countries, the drop in death rates has been followed by a fall in birth rates. The UNFPA 1981 figures give hope that this may occur in the less affluent countries. However, the causes and reasons for the decline in birth rates are not known, nor is it known whether it is a permanent decline. Were birth rates to increase in the affluent countries, no well-established theory would be upset. In place of serious well-grounded theories, there has been a great deal of unsubstantiated speculation. Some have looked to animal groups, their populations, and how they adjust to the food available to them. Others have noted that children in wealthier countries have a lesser economic value—indeed an economic disvalue—such that prospective parents are faced by a choice between having a child or a luxury, an additional car, a boat, a holiday, whereas in poorer countries they are a source of labor and lessened insecurity for parents in their old age. The greater equality of women, their greater role in deciding how many children they will have, as well as the availability of increasingly effective methods of birth control, most effectively to be used by informed, highly motivated persons in affluent countries, are also seen as being possibly relevant factors. Religious and moral beliefs are accorded some relevance by some writers, although the main stress is on economic and social causes. There is no conclusive evidence for any such theory. By contrast, surprisingly little seems to have been done by way of exploring causal accounts of variations in birth rates, and even less in empirical research to determine motivations, conscious and unconscious, that lead persons in poorer countries to have large families, and, in recent years, those in the wealthier countries to have fewer children. It is not so long ago that many who were concerned about "the ecological crisis" were predicting exponential growth of the populations of Europe and North America. Given all this, it is a good rule to suspect the most confident, brash predictions. Those most solidly based on research concerning the relevant factors are extremely tentative and qualified. The two certainties in the matter are that population, given its present age distribution, will certainly increase in the near future, and that detailed predictions of actual world population sizes at specific future dates will be found to be unreliable. It is quite possible that they will prove to be underestimates matched by underestimates of growth in food production.

There is equal uncertainty about food production in the future. There have been bizarre attempts by reference to estimates of the available agricultural hectares in the world, their productivity, and so forth, to estimate the number of persons the world can satisfactorily support. Obviously such estimates, too, are highly fallible, being based as they are on very incomplete information about the relevant factors. It is certain that there will be important technological advances in the production of food, new varieties of cereals that give greater yields, and yields in areas now unproductive. Food is likely to be derived from new sources, or from recently discovered sources, such as microorganisms. And, while it is impossible other than to guess as to the contributions that will come from genetic engineering, one need not be of a very sanguine disposition to be confident that it will make major contributions toward increasing world food production.

Given all this, and the fact that the more alarmist predictions are those based on the flimsiest premises, it might seem morally appropriate to opt for complacency and inaction in respect of the world's population growth. To do this would be wrong. One of the things for which irresponsible, unscientific alarmists about population growth are morally answerable is that their unsoundly based, unsubstantiated forecasts foster such an attitude in the minds of those who bother to examine the flimsiness of the evidence on which they are based. Against this, I suggest that, given the very great evils that would come with overpopulation, it is morally essential that mankind reflect on its moral responsibilities and rights in the matter, in particular, what are the moral considerations in favor of, and the moral costs involved in, moving toward stabilizing the world's population, at least until mankind can rationally assess the adequacy of such resources as food to sustain further growth.

The Morality of Procreation

It is useful to proceed here by noting the prevailing moral, social, and legal attitudes, and then considering the rights and duties of persons in respect of producing offspring.

Our Judeo-Christian culture has led to the development of a complex system of rules, moral, social, and legal, about permissible procreation where the context of the application of the rules has centered on the legal recognition of the monogamous marriage as the norm, and where legal sanctions apply against certain incestuous relationships and against relationships between adults and underage children, where different societies specify different age restrictions. Religious, social, and legal sanctions of various kinds have been directed against those who produce offspring outside the context of a monogamous marriage by way of sanctions against the unmarried mother and disabilities of various kinds for the illegitimate offspring, not least of which is their being accorded the status of illegitimacy or bastard. Both the legal and social sanctions directed against those who reproduce outside marriage and those who are so brought into being are lessening in many Western democracies, but some still remain (see Teichman, 1978). This brings out that there is no recognized moral, social, or legal right for consenting parties to reproduce at will. Thus, any further new restrictions on the legal and social rights would simply be additional restrictions. On the other hand, many moral restrictions have long been recognized additional to those relating to the need for the parties concerned to consent, be adult, and not be such as for their relationship to be an incestuous one.

Few moral philosophers today would seek philosophically to justify the prevailing recognized legal, social, moral restrictions on procreation. The morality of procreation is to be determined by reference to the principles of respect for persons, justice, honesty, promotion of good. Clearly, there must be respect for all parties, the actors and the beings they produce. The act of procreation carries with it the responsibility to care for the offspring. This, in turn, implies that there must be the capacity to do so, either personally or in the context of the social and welfare arrangements of one's society or state, before persons can have the moral right to bring another being into existence. The attitude expressed in encyclicals such as *Casti Connubii*, from which *Humanae Vitae* seems to be making a modest withdrawal, that it is sufficient for the married parties concerned to trust that God will provide, morally will not do. The persons concerned have no right to bring into being another person for whom they cannot themselves make provision or for whom there is no reason to believe that the state or other persons will provide. This fact imposes significant restrictions on persons. It bears on procreation by impoverished parents, more evidently but not more so by impoverished parents in an overpopulated world. And it bears on all parents collectively in a world on the edge of overpopulation. These considerations are qualified by the fact that justice, and, equally, respect for persons, for those who are so impoverished as morally to be unable to become parents require that all persons, ought to be able to exercise that basic right of personal autonomy. Justice and respect for rights dictate a reordering of the world economic community such that all but the undeserving are able equally to exercise their basic moral rights and are not denied the right to exercise them by poverty for which they are in no way responsible.

As this discussion suggests, there is and can be no abstract moral right to bring into being by a voluntary action a person who will not be cared for, who will never come to enjoy his human rights, or who, by virtue of congenital impairments, is doomed to a life of suffering. Nonetheless, there are duties of justice of ensuring that persons, due to no fault of their own, are not denied the moral right ever to reproduce. Different issues arise in determining what ought to be the legal rights of individuals. It may be the case, as liberals argue in respect of many issues, that there should be legal rights where there are no corresponding moral rights. Whether the state has the moral right to deny the impoverished person the right to have offspring will be discussed in Part III.

Morally Permissible Methods of Checking Procreation

The available range of methods of checking procreation includes sexual abstinence, abstinence from heterosexual relationships, with substitute homosexual relationships, for males, the use of institutional, sterile prostitution, sexual abstinence followed by late marriage, the use of "natural" methods of birth control, artificial methods of contraception, sterilization, and abortion.

In terms of the ethic developed in this work, sexual abstinence is a morally permissible method available to all morally responsible persons, that is, all possessed of free will. Abstinence from heterosexual relationships with substitute homosexual relationships, notwithstanding the condemnation of homosexuality by the Thomistic natural law, is also a morally permissible way of checking reproduction. In spite of the extreme range of bizarre, often inhumane proposals that alarmists have advanced as solutions to the population crisis, no major theorists appear to have advocated homosexuality as a favored method, even though there would appear to be no moral objection to such a method if it were to be effective. (For it to be effective females would have not to engage in artificial insemination.) Lesbianism alone, without corresponding male homosexuality, would be unlikely to lead to a satisfactory social order; yet there are obvious moral objections to the use of prostitution by heterosexual males; prostitution involves complete lack of respect for persons as persons.

Sexual abstinence until marriage, followed by late marriage, is also a morally permissible approach. However, when it is made to be effective by being combined with social and legal sanctions through disabilities for unmarried mothers and innocent illegitimate children, serious moral objections arise. Natural methods of birth control such as the ovulation method are morally permissible but are of limited success and utility. Artificial methods of contraception, because of their greater reliability, are clearly to be preferred, morally and in terms of efficacy. Although Roman Catholic moralists continue severely to condemn such contraception as thwarting the natural end of the sex act, their claims are to be discounted because the natural law ethic upon which their views rest is open to many objections, including its irrelevant use of "good" and its mistaken teleology. No other significant moral objections have been urged against contraception.

Sterilization raises more problems. It is widely accepted as being a morally permissible, morally desirable method of birth control, if freely entered into by the person concerned. Yet it is a kind of mutilation, if sought by the person concerned a self-mutilation of a kind to which it is questionable whether the individual has a moral right. It is a violation by the individual of his/her own bodily integrity. Such a violation by the individual may be justified if there are compelling moral reasons, but not otherwise. We have duties of stewardship in respect of our selves, our persons, our bodies. No more than others do we have a right to treat them as mere things. Sterilization is rejected by Roman Catholic moralists as thwarting the end of the sex act, and also as a form of mutilation. The same objections hold against the former claim as against the condemnation of artificial methods of contraception.

Abortion has been and is being extensively used by persons to control their reproduction. Yet, as noted in the discussion of the right to life, this is the method of birth control that morally is most open to question. There, the relevance of both the potentiality of the fetus as a person and its actuality as the kind of entity it is were noted. And, while it was conceded that the use of abortion to prevent the birth of those who would otherwise be born into misery and wretchedness in an overpopulated world could be morally permissible, morally it was more desirable that less morally questionable methods of population control be used.

In this discussion, our concern has been with methods of birth control voluntarily entered into by individual persons. The morality of the use of coercion by the state to force persons to use various of these methods will be examined in Part III.

*The Effectiveness of the Various Morally Available
Methods of Birth Control*

Of the methods noted here, voluntary abstinence and the ovulation method have a very low success rate. Those who seek to rely on them fail in their goals, and this even where there is a high degree of motivation. Such approaches could not be relied upon as methods of population control, since there would be markedly weaker motivation to realize that goal, it not being of the person's own choosing. The use of homosexuality as a means of checking birth rates has not been explored. It is unlikely to be personally and socially accepted even though morally it is much more acceptable than many of the methods involving coercion that are widely advocated. Irreversible sterilization would be effective insofar as it was accepted. It has failed to gain wide acceptance, no doubt in part because it is irreversible. Reversible sterilization, because of its reversibility, would be less effective. It too has failed to gain wide acceptance, no doubt because of the uncertainty in the concrete case whether the sterilization will be reversible. Artificial methods of contraception have a high success rate but nonetheless a significant failure rate even when used by highly motivated, well-informed persons. If they were to be adopted on a voluntary basis, not as methods of family planning but as methods of population control, they would be certain to have a much lower success rate, given that their success rate is very closely linked with the degree of motivation, as well as to the economic and social status of those who use them. Abortion has probably been the most effective method of family planning—consider its use as a backup method to contraception in the United States, the United Kingdom, and Australia, and as a primary method of birth control in Japan and countries of the Soviet bloc. Given its questionable moral status and its unpleasantness, it is unlikely that it would be readily, voluntarily accepted as a method of population control. Certainly, the motivations which lead to its use as a method of family planning would not usually be present when abortions are called for by considerations of population control.

It is for these reasons that political solutions are sought to the "problem" of controlling population growth, where the political solutions usually involve the use of the coercion of the state, economic sanctions, and the like. The morality of the use of legal, economic, and social sanctions will be discussed in the next Part.

Other Methods of Checking Population Growth

The presupposition of the foregoing discussion has been that it is only through control over or checks on birth rates that we are likely to find morally acceptable methods of controlling population growth. For the sake of completeness, other possible methods should be noted if only to be dismissed as not morally acceptable or available. Among these other methods are infanticide, senicide, genocide. Even the most extreme ecological "reformers" hold back from proposing as morally permissible the direct, intentional killing of infants, the aged, whole peoples, even though many such reformers argue that we morally ought to let die hundreds of millions of infants, adults, and aged persons in the poorer countries of the world.

The duties of the individual person. If the world is or comes to be overpopulated, the problem will be such that it can be dealt with satisfactorily only by social and political action. Individuals, acting on their own to restrict the size of their families, in a situation in which they cannot know what other persons throughout the world will do but in which they know that others are unlikely to be influenced by their example, will achieve little. What then are the moral duties of persons who come to have good reason to believe that there is a danger of overpopulation, where they personally can adequately care for all the children they choose to produce? I suggest that they adequately fulfil their duty if they have only that number of children for which they can adequately care and if they foster and support world action to stabilize the world population and to ensure that all persons enjoy the right, equally with themselves, to have offspring if they so choose. The key to the prevention of overpopulation is world political action. Action by individuals by way of limiting the size of their families, unless it is part of a general, concerted world campaign to curtail population growth, will have limited practical significance.

PART III

Ecological Politics

8

The Task of an Ecological Politics

The idea of an ecological politics may at first seem to be a strange one. Yet the discussions in Part II make it clear that if there really were to be a danger of an ecological crisis, or even of serious problems due to man's mismanagement of his affairs as they impinge on his environment, free and voluntary action by individuals, even when motivated by concern to do their duty and where their judgments about their duty are informed by the findings of ecology, will not be sufficient. There will also be a need for political action informed by ethics and ecology, as well as by the relevant social sciences, not least of which are economics and political science. Indeed, political action is basic to any real solutions to major ecologically based social problems.

The nature of the problems and the kinds of solutions that are necessary to resolve them are such as to call for a reappraisal of the sovereign nation-state and to affirm the need for an effective, ecologically informed, responsive world political authority possessed of real power. There is also a need to rethink the philosophy of liberalism, if it is to become what it must become, the basis on which these problems are to be approached. The nation-state needs to be not superseded but complemented by a world political authority. Liberalism needs to face up more fully and systematically to the need to qualify the right to liberty for the sake of respect for the rights of those now living and the well-being of those yet to be born. The greatest practical problem confronting mankind in regard to ecological problems is that of securing the international cooperation and international authority and power necessary to secure world solutions to these problems, to problems of preservation of species and wilderness, conservation of resources, supervision of science and technology, reduction of pollution, and prevention of overpopulation, while at the same time securing respect for the moral rights of all persons.

The latter consideration bears on the other dimension of the problem. Contrary to the claims of many alarmists, solutions to the ecological, environmental problems must be sought on the basis of respect for the basic values, respect for persons, justice, honesty, concern for intrinsic value as well as for human rights, where the intrinsically valuable includes pleasure, happiness, knowledge, beauty, aesthetic excellence.

Much ecological political writing exhibits two disturbing features that undermine its value. The one feature relates to the readiness of many political theorists on the basis of the flimsiest evidence to waive lightly man's claim to recognition of his human rights on the ground that the ecological crisis is so grave and critical that only the most drastic, desperate measures will succeed. The other feature is the very great confidence in the capacity of states to take whatever action is necessary to avoid the ecological crisis. In fact, the states of the world today greatly contribute to various of the ecologically based problems that confront mankind. They are unworthy of the great faith and trust that many ecological reformers appear to have in them or in the state in the abstract. A major part of the ecological political problem is that of ensuring that states carry out their ecological responsibilities. Consider here how most states contribute to wasteful depletion of resources, the production and permitting of dangerous forms of pollution, the misuse of the knowledge provided by science and the powers provided by technology, and the endangering of species and loss of wilderness. In such matters, the records of all the major countries of the world—the USSR, the US, China, Japan, France, the U.K., India, Brazil—are unimpressive. Those of the smaller countries of the world are no more impressive. Clearly, it will not do simply to trust "the state" to do what ecological political theorists tell it to do. The political problems include both determining what politically ought to be done and devising the appropriate political machinery to get done what is to be done.

These considerations need to be stressed because, while some ecologically minded political theorists propose solutions that can be implemented in and through the liberal democratic state, others seek more radical, more dangerous political solutions. Thus it is that some argue that we must abandon concern for values such as justice and respect for persons and human rights in order to secure the survival of the human race, where it seems to be seen to be a fortunate coincidence that the latter is best to be achieved by securing the survival of the society (the United States) in which the writers live. Some argue that existing liberal democratic states are inadequate, being based on shortsighted self-interest, to deal with the ecologically urgent problems. Others suggest that it is only through elitist states, socialist, communist, totalitarian of various varieties, that the appropriate goals of a resource-frugal, pollution-free, preservation-oriented, non–technologically minded society is to be achieved. Others again argue for a return to primitivism, to rural communities of the political imagination, to simple anarchical communities, in which the problems encountered in advanced technological societies are seen to disappear. Thus, the theories proposed range from the demand for the setting up of strong, effective agencies backed by new restrictive laws and creative measures within the framework of the liberal democratic state, to demands that the liberal democratic state itself be reformed and freed from its dependence on short-term self-interest, to demands for major curtailments of personal liberty and liberties for the sake of preservation, conservation, checking over-population, the control and reduction of pollution, to demands for states and/or a world political community possessed of massive, centralized powers and planning controls, where every aspect of society is planned, its economy, its production, its cities, its "wants," its "needs," human and animal reproduction, to the opposite extreme of romantic appeals for a return to the simple, unspoiled life close to nature.

Here, rather than attempting to note and examine all of such a vast range of diverse theories in detail, the various areas of ecological political concern will be considered, with a view to seeing what measures might be appropriate and necessary to cope with the problems that may arise on the basis of respect for human rights and the other basic moral values.

9

The Politics of Preservation

One of the major concerns of ecological political and moral theorists relates to the preservation of species of plants and animals and of wilderness areas, and with the state and the world political community ensuring that species are not needlessly endangered or rendered extinct, and wilderness not unjustifiably destroyed. A good deal of criticism has justly been levelled at individual states and at the world community of nations because they have failed to carry out their responsibilities in this matter, great concern being expressed about the imprudent, immoral destruction of animal species, including various whale species. In fact, it is probable that vastly more plant species are threatened than are species of animals.

In Part II it was argued that mankind has a moral duty to preserve species and wilderness unless it is very evidently in man's interest to endanger or destroy them, the duty being based on conservation of actual or potential resources and the securing of the intrinsically valuable objects of knowledge and the beautiful. On the other hand, there may be a moral right to endanger and to destroy when that is dictated by concern for human rights or human well-being. While clear in the abstract, these principles are less clear in their practical applications because of our limited knowledge, indeed, ignorance, concerning what may become a resource in the future. A species that today is a nuisance or worse may in the future prove to be of great value. Since preservationists do not seek to preserve disease organisms or parasites which use human beings as their hosts, and since the species that are endangered by deliberate overkilling, as with whale and fish species, usually are such that they need not and ought not so to be endangered, the difficult cases that arise in respect of justifiable eradication of species are relatively few in respect of intentionally threatened species. The more difficult cases relate to the use of wilderness for the production of resources, food, timber, extraction of fossil fuels, uranium, and other materials, where the encroachment on wilderness, besides constituting a loss of wilderness, threatens rare species of plants and animals.

It will be argued here that the state has the right and duty to preserve species and wilderness when this is not contrary to respect for human rights and well-being, or when it is dictated by concern for preserving the intrinsically valuable, but that it has no right to preserve species and wilderness simply for their own sakes when they possess no intrinsic value and in no significant way contribute toward human well-being, and when concern for human rights and well-being or for the intrinsically valuable would dictate not preserving them. Guidelines that might seem to be useful to states are that preservation of species is distinct from preservation of all the members of a species, the former being compatible with restricting the numbers of the members of a species, or confining it to localized areas, and the like; that there is a presumption in favor of preservation of species, such that a telling case needs to be made out before it is permitted that a species be endangered. There is not the same presumption or such a strong presumption in respect of preservation of wilderness where encroachment does not endanger species. If land needed for food production is suitable for agriculture, and if persons will starve if the land is not used, there would appear to be a clear case for using the land. So too if the land contains an essential scarce resource, one needed if society is to function without economic disruption, if persons are to have employment and the goods and dignity that come with that. Yet so to exploit wilderness may be to endanger species. Thus, in its concern for preservation, the state must determine what considerations morally justify overriding any presumption there may be in favor of preservation. It must also be vigilant in determining when human action and inaction come to create a danger of a species being endangered. It will need to determine what are the most effective means of securing that preservation that ought to be secured, the use of criminal law, the setting up of preservation agencies, the conferring of legal rights on natural phenomena, legal personhood, the permitting of class actions, or the like. It will also need to ensure that its coercive powers are not wrongfully used to curtail human freedom, to deny human rights, and lessen human well-being, simply for the sake of preserving species qua species, wilderness qua wilderness, when such preservation has no justification, and this for the reasons indicated in Part II.

Preservation Dictated by Concern for Human Well-Being, Human Rights, and the Intrinsically Valuable

States, independently and collectively, as members of the world political community, must determine what species and wilderness should be preserved, and, when species and wilderness are wrongfully threatened, what are the appropriate means of securing that preservation that ought to be secured.

As noted above, the task of determining when to permit and when to oppose the endangering of a species will rarely be difficult when it is the deliberate eradication of a species of plant or animal that is being considered. It will be rare with organisms other than disease organisms and parasites that a strong enough case could be made out. Usually it is sufficient for man to check the numbers of the species he sees to be pests, where many of these are species introduced to new countries and new environments. For many states, the decision has already been taken in respect of wilderness. They possess no wilderness areas. Those that do possess wilderness areas will face increasingly difficult decisions. As the world's population increases, and the justice of the demand that all persons have access to a reasonable standard of living comes widely to be accepted, there will be increasing pressure for wilderness areas to be opened up for cultivation, for the resources they contain and the resources they can be used to produce. Consider here the wilderness areas of Australia and the desperate plight of hundreds of millions of persons in Asia. With the encroachments on wilderness will come serious threats to species of plants and animals. It is in judging what is an adequate justification for encroaching on a wilderness area for mining, timber, agriculture, recreational uses, and the like that states face their major decisions about preservation. It is in respect of these decisions that they are making very serious mistakes, often by way of underestimating the case for preservation. Nonetheless it does need to be acknowledged that many of these decisions are very difficult ones, involving as they do the resolving of clashes of values in contexts in which there is a great deal of uncertainty, even ignorance, as to what will be used or be usable as resources in the future.

The moral decisions are difficult enough even when made with complete honesty and disinterestedness. In fact, often the context is one in which considerations of political expediency may make a completely disinterested, honest judgment difficult. Further, often vast sums of money hang on how a decision relating to preservation is resolved. Politicians, and the bureaucrats who supply them with the relevant information (and perhaps misinformation) and advice, may be subject to pressures and inducements ranging from indirect offers of assistance to outright bribes. Even where outright corruption does not occur, there may be more insidious, less obvious corruption.

Political leaders in Australia, and I should expect elsewhere, regularly betray the people they represent by engaging in seeming economies, by alienating parks and public gardens for the building of hospitals, schools, freeways, even prisons, thereby keeping the seeming costs of projects demanded by the public to acceptable figures, at the cost to the public, those now living and those yet to be born, of the loss of irreplacable park land. The public, present and future, is tricked into bearing that cost. Wilderness is likely to be sacrificed in the same way, and on the basis of the same shallow thinking. It will often prove cheaper to use wilderness areas than to buy areas that are already developed—cheaper, that is, in seeming money terms, not in terms of real costs. This bears on the difficulty of estimating the value to man of wilderness areas. The Great Barrier Reef is one of the finest, most magnificent wilderness areas in the world. Yet, if as may prove to be the case, vast oil reserves are located in that area, there will be tremendous pressure on political leaders from the value of such a resource, from the political advantages of developing it in terms of the economy, political advantage which will result therefrom, and from those who stand financially to gain from developing the area. In such a context, those who will wish to preserve the wilderness will be confronted with the problem of establishing that its intrinsic value justifies forgoing all the benefits that would come with development. That is a hard enough case to make out. In the concrete political situation, given the partiality of politicians for short-term expediency at the cost of long-term goods, and given the vulnerability to corruption of politicians and bureaucrats even in the most enlightened democratic states, for preservationists to make the case successfully will prove vastly more difficult. That is why concern for preservation cannot safely be left to politicians and bureaucrats to decide case by case, such that political expediency, lack of knowledge, public interest or lack of it, may be decisive. Other effective appropriate methods of securing that preservation that ought to be secured are needed.

Clearly, states can and ought to use noncoercive approaches to promote a concern for preservation, approaches through education, the dissemination of information, the use of tax incentives, subsidies where desirable, and the like. While such methods will have only a limited value, they are nonetheless important, worthwhile approaches, not least because they will contribute to there being a well-informed, concerned public opinion that will constitute a very important back-up for coercive measures and for keeping the state itself preservation-conscious. Community awareness, community vigilance, and a sense of community responsibility are of great value, especially if the education comes to inculcate a sense of moral concern by individuals that they as individual persons have a moral interest in preservation. That will greatly reduce the threat to species and wilderness from ordinary persons. It will have less effect on professional killers of game, collectors of flora, mining companies, and the like, but it will set up a climate of opinion in which their activities are viewed much less favorably.

The use of coercive measures is also necessary. This can take the form of blanket ~~laws, with or~~ without exceptions written into them, or protecting specific species or specific wilderness areas. Various animal species, aquatic and land, may need only limited protection, for example, for only part of the year. Plant species may need protection only in specific areas, and not wherever they occur, and so on. It may be necessary to restrict the property rights of owners of wilderness areas. If so, compensation or other arrangements may be appropriate. Such laws, if enforced and policed, will serve to secure preservation against the inroads of private persons and corporations, provided that they are not repealed at the behest of those who wish to exploit the species or the wilderness area. However, species and wilderness can be endangered accidentally, inadvertently, even unknowingly, as when heavy machinery carries a disease from one area to another in logging, road making, mining exploration, or when new species are accidentally or deliberately introduced into an area and they become predators on a rare species or on the organisms on which it feeds. The most effective way of securing wilderness and the rare species it contains—and it is not a sure way—is to ban entry to it. Even entry on foot, of the kind desired by so many exponents of preservation of wilderness and species, can drastically change the ecosystem by introducing destructive organisms.

The big problems today are those posed by the state itself, the world community of nations, and the big, usually multinational, corporations. The first and last can be checked by states being encouraged to do what many have done to a small degree, namely, to set up independent, well-financed, strongly empowered preservation agencies, which have the resources, financial and otherwise, to investigate whether undesirable destruction of species or wilderness is occurring or is threatened in the future, and to take strong action to prevent it. Such agencies would need to have the power to act against the state as well as against corporations and individual persons and would need security against reprisals by states curtailing their powers, cutting off their funds, and the like. A preservation authority would need to be able to take the initiative to have unprotected species protected as the need arose. It would have the roles of both determining when protection was needed and of providing the protection. It would therefore need to be accessible to the community, to interested groups, to those with an interest in preservation and those with an interest in not preserving. There would be serious risks of inadequate preservation measures if such an authority had limited authority and limited resources. There would be also serious dangers to the community if such a preservation authority could always override the judgments of governments, corporations, and individuals without appeal. Its judgments could be mistaken, its priorities could be wrong. It could frustrate very desirable, even socially essential measures. It is quite likely that had there been such a preservation authority in Australia when that very important national development, the Snowy Mountains Hydroelectric Scheme, (the Snowy Scheme) was adopted—a scheme that involved great interferences with wilderness areas, redirection of rivers, creation of reservoirs, a building of complex road systems, hydroelectric power stations, and the extension of irrigation systems hundreds of miles from the main area of activity—it might well have been approached to oppose this scheme, and might have done so. Hence there must be a right of appeal by individuals, corporations, and the state itself to the courts, against the rulings of such an authority. This means that ultimate power, ultimate safeguards, would be in the hands of judges.

Other approaches, which could be complementary to the foregoing, are those already advocated by many preservationists in terms of extending the range and scope of class actions to permit injunctions to prevent interferences with nature, where interested parties are given standing, actions in the public domain, and the according of legal rights to natural objects. Typically, class actions are directed at securing payment of damages, but they are also used and are usable to secure injunctions. They, and actions in the public domain, while in practical terms very like actions to protect legal rights conferred on natural objects, would have a different logical basis. The former (class actions) would rest on protecting existing legal rights of persons. With actions of the latter kind persons would come to have legal rights as representatives or guardians by virtue of the legal rights of natural phenomena.

There is much to be said for and against such approaches. Consider the approach by way of class actions. Class actions, simply by causing very costly delays, even when not legally successful, could effectively block very desirable developments, vital developments, by increasing costs beyond that which the industry could or would afford to bear. And they could be successful when they ought not to succeed. Again, consider schemes of the importance and magnitude of the Snowy Scheme. Had class actions been possible when that scheme was initiated, and had the public been as preservation-minded as it now is, and had the ordinary citizens, or those who were parents and hence with an interest in the future of Australia, its wilderness, fauna, and flora, had legal standing, that scheme could well have been thwarted. More generally, class actions give power without responsibility, such that vexatious and malicious litigants, who in a welfare state would have their legal expenses paid by the state, could exercise great power without responsibility to anyone, given that delays can greatly increase costs and cripple important developments. Similar problems can arise in respect of actions in the public domain. Such nonresponsible power is not to be conferred lightly on persons and widely distributed throughout the community. If it were to be so distributed, preservation class actions could open the way to nuisance actions, to action by political agitators, malcontents who use the courts to disrupt economic and social planning vital to the stability and well-being of the community. Further, this approach again confers on the courts, more immediately than does the use of preservation agencies, great power to determine what innovations affecting the environment are permissible in a period of great techno-logical innovation and change. Yet, in most societies, the legal profes-sion, and most evidently the judiciary, are among the more conserva-tive elements. Nonetheless, with appropriate safeguards regarding who is given standing to initiate class actions for the sake of preserva-tion, this kind of approach may have a useful role to play.

As this discussion brings out, there are dangers in not preserving, and there are also dangers in preserving at the expense of develop-ments vital to the well-being of the community and its members. A very delicate balance of powers and checks is needed.

The State and Preservation for Its Own Sake

In Part II, it was argued that not all wilderness, and not all species, are intrinsically valuable, and hence that there was no moral case for preservation from the intrinsic value of all species and wilderness. Yet many, including some who acknowledge this, seek to argue that species and wilderness ought to be preserved for their own sakes and that the state has a right and a duty to preserve wilderness and species for their own sakes, and this at the expense of human rights and human well-being. Preservationists are meeting with a disturb-ing measure of success in promoting this view. Yet such intolerance by the state is politically and morally indefensible, and beyond the proper exercise of the state's authority. This is so whether or not the action is directly coercive, as in the use of the criminal law, or primarily noncoercive, by means of education, propaganda, and coercion through the use of funds raised by taxation to finance these noncoercive measures. Laws directed at protecting venomous snakes by making it a crime to kill them would seem to illustrate this concern for preservation's sake, in that while it may prove that snakes will come to be a resource—as would be the case if important medical uses were to be found for their venom—on the weight of evidence today they are a nuisance and a danger to human life and health, concern for which dictates that individuals be accorded the freedom to kill such snakes, to rid their properties of these pests even at the risk of rendering them endangered species. (Sufficient specimens could be retained in zoos throughout the world, where they survive well, so as to make the risk of any species of snake becoming extinct very slight.) To date there are few restrictions making it a crime for

individual persons to invade wilderness areas for recreational purposes or to convert wilderness areas they own to domestic or commercial use, and this no doubt because wilderness lovers love to
trespass on and to own bits of wilderness and have been less active in
this area. The danger nonetheless is there. A case needs to be made
out for curtailing the exercise of human rights and for acting against
the interests of human well-being.

As was noted in Part II no preservationist seems prepared to adopt
an absolutist moral position. Yet, if preservation is opted for for its
own sake and not because and insofar as wilderness and species
possess intrinsic value, it is impossible to determine what sacrifices of
human life, human health, and human liberty are to be seen as
justified by preserving wilderness and species, and hence precisely
what intolerance of the exercise of their human rights by persons
would be favored for the sake of securing the survival of a species of
reptile, fern, moss, fungus, or one hundred hectares of wilderness.

The politically intolerant preservationist demands that the state act
intolerantly, that it impose his/her favored value judgments/prefer-
ences on the community at large, even when so to act is to act
contrary to the well-being of the community, even of the world
community. It is to demand the kind of intolerance that Mill's
celebrated infallibility argument is directed at condemning as unjustified. It is intolerance that could be justified only if the moral judgments and preferences underlying it were infallibly known to be true.
As noted in Part II, far from being infallibly known to be true, no
sound basis of any kind has yet been advanced in their support. At
best they are simply fallible opinions that all species and all wilderness prima facie ought to be preserved. More evidently, they appear
not to be fallible judgments but mere preferences. At worst, they are
false beliefs about the intrinsically obligatory or the intrinsically
valuable. Certainly it is not an infallible truth that it is more important
to preserve venomous snakes than to save the lives of the many
Australians who die each year from snake bites, and to save others
from distress, sickness, fright, inconvenience, and nuisance caused
by snakes. The preservationist's moral attitude is not one that com-
mends itself as that to be accepted by rational, benevolent men of
good will. It is self-evidently better to save the child from death from
snake bite by killing the snake that threatens it, whether or not it be a
member of an endangered species, than to save the snake. How then
can the state be entitled to curtail human rights and sacrifice human
well-being on the basis of such a moral view or preference of the
preservationist?

The important argument for liberty from human self-development
and its value, and from liberty being a condition of self-development,
also undercuts the preservationist's demand for intolerance of those
who endanger wilderness and species for man's good and man's self-
development. The preservationist's intolerance would deny to many

the opportunity to be effectively self-developing. Also relevant is the argument from respect for persons and for the autonomy persons possess. What moral right can another person have to use the state to thwart the self-determination of a rationally autonomous person, in order to impose his highly fallible moral judgments, or his mere preference?

That the adoption by states of intolerance for preservation for preservation's sake violates not simply the rights to liberty, self-development, and respect as persons but also to life and health further emphasizes the radical, indefensible, undefended nature of the preservationist's intolerance.

Let us now consider what arguments might seem to be available to the preservationist. The only considerations that the preservationist seems to be able to offer are negative ones, that liberalism, including the most recent liberal theories of natural, human rights, are human-centered, and, as such, false. Clearly, this will not do. Such a claim amounts simply to a characterization of the arguments, not a reply to them. Argument is needed to show that, when and insofar as liberal theories of rights are human-centered, they are false. Further, a justification of the kind of political intolerance preservationists favor is not possible, and this for the following additional reasons.

It may be argued that the foregoing argument leans very heavily on the moral argument that intrinsic value cannot properly be ascribed to species qua species, wilderness qua wilderness, but only to certain species and to certain areas of wilderness, that if it could really be shown by the preservationist that species and wilderness have intrinsic value, the above argument would fail. The preservationist cannot establish the intrinsic value of all species and wilderness. However, even if he could make out a telling case for his view, the arguments for liberty, most especially the infallibility argument, would still undercut his case for intolerance. Even without these arguments being considered, the preservationist would still need to set out his moral criteria, guidelines for weighing the intrinsic value of species and wilderness against other intrinsic values and against human rights, human life, human well-being. (It is unclear whether such preservationists see all species, all wilderness areas, as of equal value, and a species to be of the same value as, say, a hectare of wilderness.) This would lead to a much more qualified defense of preservation. However, the preservationist cannot establish the intrinsic value of species and wilderness in the relevant sense.

Therefore, to justify his demand for intolerance, he has to do two things. He has to show that there are good reasons for rejecting the arguments for the right to liberty. And he needs to show not simply that there are good reasons for us as individuals to accept, act upon, and enforce conformity with his moral attitudes or preferences (or intuitions if he claims intuitively to discern intrinsic value or intrinsic obligatoriness), but that these reasons apply to the state as well. This he is evidently unable to do. He must develop a whole new political theory of the role of the state and of its use of coercive power. It must be one that allows as legitimate his kind of intolerance to enforce respect for species and wilderness, but such as to provide no basis for an extended intolerance in respect of the enforcement of moral, religious, and other beliefs. This is because such a preservationist is usually opposed to Christian and other religious beliefs about man and his relationship with nature and therefore would not wish to have such beliefs imposed by and acted upon by the state, as being true beliefs. This too is an extremely difficult, philosophically novel enterprise. The arguments that moral and religious paternalists such as Plato, Aristotle, Saint Thomas Aquinas, and Pope Leo XIII, offered for moral and religious intolerance are much more plausible than anything that the preservationist has to offer. These paternalists were concerned to protect the individual from error and evil, from moral harm, harm to man's immortal soul, to prevent evil, wicked, mistaken persons from corrupting and harming others. They believed that the individual is not the best judge and guardian of his own moral and spiritual interests, and that, without protection, he would not find or retain true belief, or enjoy the good life. These arguments for intolerance are of no value to the intolerant preservationist, being as they are, human- and/ or God-centered.

The liberal rejects the claims of this kind of paternal authoritarianism on the basis of the arguments from the need for infallible knowledge, the lack of certainty about value and other judgments, the argument that we are better judges and guardians of our own interests than are paternal authoritarians, except in certain characterizable kinds of cases, that morality involves the exercise of our moral autonomy, that self-development is a major good and its pursuit an important part of being good persons and good as a person.

If the preservationist insists on his right to impose his moral preferences and priorities on the community, he must give up the infallibility argument in toto, or satisfactorily explain why his views should be accepted as exceptions to it. Yet, in spite of its many qualifications, the infallibility argument is a powerful argument for liberty. Similarly, the preservationist will have to reassess the significance of the other arguments for liberty. This suggests, as is the case, that the task confronting the preservationist authoritarian is an impossible one. That we do not in fact confine ourselves to concern with purely human goods and evils, but value rationality, pleasure, happiness, beauty, wherever they occur, and disvalue irrationality, stupidity, pain, suffering, ugliness, wherever they occur, shows that the preservationist's critique of liberal ethics and politics as wholly human-centered is mistaken. What is true is that if intrinsic value could be shown to inhere in wilderness and species as such, the liberal would be concerned that full moral and political note be taken of this fact, as he already does in respect of natural beauty.

It may be replied to the above that this shows that it is wrong for a state to use its coercive powers for preservation for its own sake, but that it is legitimate for the state to foster preservation by noncoercive means. Here, too, a justification is needed, however. What possible reasons could a state offer for using its powers and funds for noncoercive action for the sake of preservation as an end in itself? Even to have the beginnings of a justification of any kind, a state would need to argue that preservation was worthwhile for its own sake even when disadvantageous to man. Otherwise it would be indefensible for the state to use its authority, its manpower, its funds, raised by coercive taxation, to preserve what intrinsically had no value and what instrumentally is disadvantageous to man. Such noncoercive support for preservation would be politically wrong and mischievous, in the same way it would be wrong for a state noncoercively to support a religious organization that caused harm or which had no serious claim to truth or goodness. Noncoercive aid can no more be justified than can coercive aid for preservation for its own sake.

Having argued in this way against political action for the sake of preservation as an end in itself, it needs again to be stressed that political action for preservation for the sake of human well-being, human rights, and to preserve what is intrinsically valuable is of the greatest importance. The interconnectedness of the world ecosystem makes it dangerous to endanger or seriously reduce the number of species, and the number of members of many species without good reason. So to act is to fail to conserve what is or may prove to be a valuable resource for man. Further, much of nature is beautiful and such as to give pleasure to those who behold it. It is to be preserved if possible for these reasons. Nonetheless, it remains true that there are and will continue to be important clashes involving basic values, the need for agricultural land for food production, timber for housing and paper, the need to enter wilderness areas to mine scarce resources, where the wilderness may be damaged as a habitat for rare species of plants and animals, and so on. Further, it is presently impossible to engage effectively in efficient food production without the large-scale use of pesticides and fertilizers. We cannot adequately protect human health without using pesticides, antibiotics, and other biocides. We need to interfere with rivers to prevent floods, to provide irrigation water, and so forth. We cannot afford not to reclaim land that can be made agriculturally productive in the way that significant areas of the Netherlands and the Wash of England, and in less recent times, so many areas in Europe, Asia, and North America, have been reclaimed and made productive, and this without serious ecological harm. (Of course, when considering reclaiming land, possible adverse ecological effects need to be investigated before action is taken.) We need to exploit wilderness areas for fossil fuels and other key resources. In all such interventions, the wise, informed, ecologically minded state will take heed of possible dangers that come from upsetting ecosystems, endangering species, and losing to the world organisms valuable to man.

10

Animal Rights?
Political Implications Thereof

In Part II it was argued that, in spite of the claims to the contrary by many environmentalists and animal rightists, animals that lack autonomy do not and cannot possess moral rights. However, given the increasing acceptance of the view that sentient animals do possess moral rights, it is worth considering the political implications of such a view. Possession of moral rights, whether by animals or humans, is clearly a fact of which the ecological state would have to take full account.

With human moral rights, typically the possession of a moral right constitutes a ground for conferring legal rights to freedom from interference, and legal rights to those aids, facilities, and conditions dictated by the right. This does not apply in all cases, as, for example, when the exercise of his right by the possessor of the right will violate an equally or more stringent right of another person or persons. Following this model in respect of animal rights, it would seem that a moral state would be obliged to protect animal rights on the same bases, and in ways similar to those in which human moral rights are protected. The big difference would seem to be that animals must be represented, have guardians to seek to secure for them the enjoyment of their rights and also paternalistically to ensure that they do not exercise them to their own detriment if that supervision itself be compatible with respect for their rights.

The rights most likely to be ascribed to animals are the rights to life, health, bodily integrity, liberty, and freedom from suffering, and perhaps also the right to development of their beings. Protection of animal rights against human violators of rights is fairly straightforward. It would involve making it a crime for humans to invade rights by actions of commission and omission. It would also involve the state in setting up animal health services to which owners of animals—if ownership were not itself to violate the animal's rights—were obliged to bring sick and suffering animals. It would involve making eating meat and using leather (other than from animals that die naturally) a crime, vegetarianism legally obligatory. Dogs, cats and farm animals would have to be freed from the physical constraints now imposed on them except where dictated by concern for their rights. And the lives of rodents and other pests would have morally prima facie to be respected. All this would very seriously curtail human liberty in many ways. The measures would have other momentous, far-reaching consequences, initially in lessening the supply of food available, and then, depending on animals' rights to breed, or our right to restrict this, permanently reducing the food available or making very much more vegetable food available. Pressure would be put on industry and technology to provide substitutes for animal products such as leather, fat, fur, and possibly wool (depending on whether it was a violation of a sheep's right to shear it).

The protection of animals in the enjoyment of their rights against other animals would create massive theoretical and practical political problems. The state would be obliged to protect rabbits and other prey from foxes, and yet not bring about the death of foxes; protect birds, rats, and mice from cats and dogs, but keep cats alive; and similarly with lions, tigers, and other animals of prey, without killing other animals to feed them. If fish are among the possessors of moral rights, the same would hold for them. There, the practical problems would be immense. These considerations may lead to the suggestion that the state is morally obliged to protect animals only from human interference. Yet, if we consider human moral rights as they hold for infants and others who cannot protect themselves from natural forces and dangers, this seems not to be what follows from the possession of a full-blooded moral right. Helpless animals who possessed moral rights would need protection similar to that needed by helpless infants.

The implications of the recognition of animal rights for criminal law, the punishing of those who kill or do not help dying animals in their care, who are cruel to animals, the enforcement of vegetarianism, bring out how artificial is the distinction drawn by many liberals between the right of the state to act to protect moral rights and the state's lack of a moral right to enforce morality. What is true in respect of the legal protection of human rights by the use of criminal law is very evidently true in respect of similar protection of animal rights. It amounts to nothing short of the enforcement of morality. What is true of the ecological state in this matter, if animals were to be shown to possess moral rights and the ecological state protected them, is also true of enforcement of concern for preservation of species and wilderness, whether or not it be for human well-being, human rights, preservation of the intrinsically valuable, or as an end in itself; it would also be true of the enforcement of conservation of resources generally, of conformity with antipollution measures and laws directed at checking population growth, if these measures were morally justified and morally justifiable.

11

Resource Liberality Versus Resource Profligacy or Resource Frugality

It was urged in Part II that the case for a general policy of conservation of resources is morally a weak one, more especially because it inevitably as a practical political reality would be at the cost of securing the well-being and respect for the rights of persons in the Third World. There is a stronger moral case for selective conservation, where the case for conservation is assessed resource by resource. There is an unanswerable moral case for conservation of renewable resources such as land, sea, and air, keeping them renewable. The considerations noted in that discussion were: (1) Nonrenewable resources, if used at all, must ultimately be exhausted; that is unavoidable. Hence a policy of conservation, if it is not to be thought to be contributing to bringing about an even greater ultimate catastrophe, must rest on a belief that man's technological ingenuity will, if given the additional time that conservation will provide, find solutions to the problems posed by the ultimate, unavoidable exhaustion of nonrenewable resources. (2) We lack knowledge about future technological developments and about the resource needs of future generations, and, indeed, about their numbers, population policies, styles of life, and the like. (3) We cannot know how much conservation will be advantageous to man. Too little may deprive us of important goods and yet achieve naught. Too much will lead to a waste of resources that could have been used to great advantage. (4) The resources that it is proposed to conserve are needed today by the impoverished countries of the world. It is a practical, political reality that conservation as a policy today will be at the expense of raising the living standards of the needy billion or more persons in the Third World. This is acknowledged explicitly by many who advocate a policy of conservation of resources. (5) Determinate duties to existing persons, to secure them the enjoyment of their basic rights, override speculative duties to possible future persons.

It is also a political reality that a very considerable number of ecologically minded economists, moralists, and political theorists continue militantly to advocate the adoption by the Western world in general, and the United States in particular, of a policy of conservation, this commonly being defended under the inaccurate demand for "a no-growth society." Their campaign, while meeting with no significant success in influencing the policies of governments and large corporations, is meeting with surprising, although modest, success in attracting the attention of academics, and even converting many to join their ranks. It is therefore desirable that consideration be given here to the politics of conservation. Questions that need to be answered include: What kind of economic, social, and political measures would be necessary to implement a policy of conservation of resources? How is such a policy to be brought into being? Can it successfully be implemented within the framework of a liberal democratic political order? What would be the costs of opting for such a policy in terms of freedom, democracy, enjoyment of other basic rights, and the consequences not simply for members of the affluent countries but for the majority of mankind, the members of the Third World? Finally, what are the alternatives, their benefits and costs?

The Political Problems of Conservation

The first reality that needs to be faced if there is serious concern about conservation of resources is that it is a world issue, which to be implemented successfully as a policy requires world action—concerted, cooperative action, desirably by all nations, but at the very least by all the major nations, major in the sense of resource use and/ or population, in the world. There is little or no point in one or few nations opting for a policy of conservation of resources, if other nations continue on a policy of profligacy in resource use. Resource-rich states may cushion the effects of a world resource crisis by opting themselves for a policy of conservation, but unless they are completely self-sufficient in terms of all key nonrenewable resources, they will simply cushion the crisis, as they would be obliged, as the crisis loomed, to trade their conserved resources for those they lacked. Further, their very policy of conservation would put them at risk of attack by resource-starved nations if they refused to trade and thereby erode their conserved resources. Consider what would happen if the oil-rich nations of the world were to choose a serious policy of conservation of petroleum, producing and making available only a quarter, a tenth, or less, of what they now produce. It is true that if a high-resource-use country such as the United States opted for resource frugality, this would have a significant effect on resource depletion. However, unless the countries of Europe, the Societ bloc, Japan, and even such countries with much smaller total resource use as Australia also opted for conservation, the problem would remain. The most vocal exponents of the no-growth society give little or no attention to this very major aspect of the problem: the need, if the policy is worth adopting at all, for it to be adopted as a world policy.

The importance of this is confirmed by two further considerations. A very important area of waste of key resources is in what is called "defense expenditure," preparations for war. If all the primary and secondary "defense" uses of resources by all the nations of the world could be computed, it would come to an immense figure. Unless there were to be developed a world conservation policy that terminated such wasteful resource use, whatever happened in other areas of their economy, nations would have to continue to direct resources, and continually increase the resources so directed, to defense measures, or leave themselves vulnerable to aggressor nations. Given that the adoption of a no-growth economy is likely to lead to resistance to technological innovation, and that part of the reason for the public accepting the massive costs of defense research and production is that it so often brings real benefits in peacetime to the ordinary citizen (consider the many benefits that have come with the space exploration program), one does not need to be pessimistic to believe that if the United States and Western Europe unilaterally opted for a no-growth economy other than in the area of defense, it would ultimately have a spillover effect on defense, leaving them ill-equipped to deal with the likely aggression of the Soviet-bloc nations. Wittingly or unwittingly, the advocates of no-growth, unless they set out real and adequate safeguards for defense research, expenditure, and resource use, are advocating a course that must make the free world vulnerable to aggression. Yet, if resource liberality is sanctioned in that area, it seriously undercuts the purpose of the policy.

A world approach is also necessary because a serious conservation policy dictates that the multinational corporations who are themselves profligate in their use of scarce resources, and which for reasons of profit are not averse to rendering renewable resources nonrenewable, and which foster trivial, wasteful uses of resources, must be controlled and made answerable to the world community of nations. Their vast wealth, their great power, and their amorality make it extremely difficult for individual nations to control them. Yet, if there is to be conservation of resources, they must be controlled and made answerable to the community of the world. They now have more power, more wealth, than many nations, but they are fully answerable, fully accountable, to no one. This need for world control of multinational corporations relates not solely to issues of conservation. A much stronger case can be made from concern about pollution and preservation.

Most exponents of a policy of conservation ignore the fact that if it is to be a serious policy it must be embarked upon by the world community of nations. Instead, they concentrate on discussing it as a problem to be tackled by an individual nation. Since most advocates of this policy seem to be citizens of the United States, it is in the context of that nation and countries under the U.S. sphere of influence that their expositions of the policy occur. Usually they use the terminology of a conservation-based society being a no-growth society, when in fact what they mean to support is the very different concept, that of a resource-frugal society.

It is a gross confusion to equate a no-growth society and a resource-frugal state. A resources crisis would dictate simply resource frugality, and no-growth only to the extent that that is dictated by resource frugality. Significant growth using renewable resources is compatible with a nonrenewable-resource-frugal economy. This would much more evidently be the case if natural energy sources were to be harnessed effectively. If solar energy, winds, tides, geo-thermal energy, and the like could be used extensively in conjunction with renewable resources, a good deal of growth would be possible. In any case, many things that contribute to economic growth are low in resource use, not the least important of these being amusement activities and education in the liberal arts. By contrast, a no-growth economy, even in a depression, can be a high-resource-use economy. It was during the Great Depression that such a valuable resource—land—was squandered by overuse and undercare. The confusion of the two concepts, that of a resource-frugal economy and of a no-growth economy, is inexcusable, but significant. It would appear to spring from the ascetic, puritan moral stances of exponents of the no-growth economy, they favoring no-growth on independent moral grounds and seeking to buttress their moral preferences with economic supports.

That a resource-frugal economy need not be a no-growth economy is not unimportant. While both types of economy and the societies in which they operate would require a great deal of centralized planning, there would be room for more innovation, mobility, and increased standard of living in a resource-frugal society.

The resource-frugal society could be realized in various ways, one being by way of rationing scarce resources, having "depletion quotas" or the like, and another by having a centrally planned economy based on low resources use, efficient production, and elimination of built-in obsolescence and other wasteful practices, with production according to real needs and "legitimate wants."

For the approach by way of rationing of resources to operate without bringing in its train not simply a lower standard of living but positive hardship, the quotas of resources permitted to be used would have to be flexible and related to population growth, or, if they were to be fixed for all time until the various nonrenewable resources are exhausted, they would have to be backed by measures to control, indeed, to prevent, population growth. Flexible, upwardly increasable quotas would mean less conservation of resources than many exponents of the resource-frugal society deem desirable. With significant population growth, it could permit considerable increases rather than decreases in resource use. Yet, for reasons to be developed in chapter 14, there are grave reasons for doubting whether the morally acceptable methods of controlling population growth will either be accepted, or, if accepted, will work to the degree necessary to underpin a resource-frugal economy founded on fixed, unchanging resource quotas related to a fixed, stable population.

A second problem relates to the basis for determining the quota, the amount of the resource to be made available each year, or each decade. On what basis, in terms of what guidelines, could a state rationally determine how much petroleum, coal, mercury, iron, aluminum, and the like to make available for use? Would it reach its decisions on the basis of present use, present reserves, the rate of use by other nations in the world? Would it revise its quotas in the light of the discovery of new reserves, new resources, the falling in demand for the products that use the resource? Fixed quotas, and quotas that were adjusted solely on the basis of population changes, could still go hand in hand with great waste, with built-in obsolescence, high-consumption cars, high-energy-consumption plants and homes, the poor being priced out of the consumer market.

If the grossly inegalitarian consequences of leaving it to the market to determine who acquired the resources and the inefficiencies and notable wastefulness of the ensuing production were to be avoided, the quota of the resource would need to be distributed to producers by central planners who made judgments about relative priorities of products. This would make the rationing approach come very close to the approach of having a central-planning state.

Further, an effect of such rationing of resources, most evidently but not only of energy resources, would be greatly to reduce production and thus employment. Although a low-resource-use economy need not be a no-growth economy, it would be one in which there was much less work to be done than in a resource-profligate economy. Hence, either vast numbers would be unemployed and destitute, or work would be shared thinly throughout the community with relatively few hours being worked by each worker; in the latter case, either the wages paid would be inadequate to sustain the worker and his dependents at an adequate level, or they would be made vastly higher on an hourly basis than wages now are, with greatly increased costs of production. Obviously, the course of having a large, destitute class of unemployed would be morally intolerable. Yet, to provide adequate welfare benefits would again be to greatly increase costs of production. The only morally viable society is one in which those who are unemployed due to the system and not to their own unwillingness to work have the status of respected members of society, accorded not a dole but an income adequate for a meaningful, dignified life. Given all this, it is hard to see how a resource-frugal economy based on a rationing of resources could fail to lower real standards of living significantly.

A centrally planned, resource-frugal economy which eliminated wasteful practices such as built-in obsolescence, production for artificially created "needs," wasteful uses of energy caused by reducing the charge per unit the more energy used, and the like is, in principle, compatible with a real and even with an economically measurable increase in the standard of living, this of course depending on the degree of resource frugality for which the society opted. If more durable goods were produced, if luxuries that are desired only due to pressure advertising were dispensed with, if wasteful uses of energy were no longer to be encouraged, and so on, the resources now used in their production could be used to meet other, more serious demands.

The major difficulty in filling out the picture of the resource-frugal, centrally planned society, a difficulty not satisfactorily resolved in discussions such as those in D. C. Pirages' *The Sustainable Society* and M. Olson and H. L. Landsberg's *The No-Growth Society* and similar works, relates to the criteria to be used in determining what is to be produced in such a society. "Production according to needs" sounds informative and impressive, yet in any straightforward, philosophically useful sense of *need*, such production would be unacceptably low, permitting much inferior housing, amenities, and food to that which is now reasonably demanded. This principle might be amended to production according to real needs and "legitimate wants." People demand clothing, housing, furniture, transport, appliances, energy, amusements, greatly in excess of need. A resource-frugal, centrally planned state would have to determine what are real needs and legitimate wants and relate production and resource use to them, at the same time ensuring that there was no overuse of resources. It would decide such questions as whether any, some, or many ordinary citizens were to be permitted to own cars, and if so who; the size of houses and the facilities they could possess; the heating and cooling systems if any; the energy-using appliances that could be manufactured and used; the restrictions to be placed on energy-consuming facilities such as TV; the rationing of newsprint to newspapers and publishers; and so on. On many issues, such as the keeping of domestic pets, cats, dogs, and birds, a total ban would seem likely.

No objective standards exist for determining whether a demand for a car, several cars, a TV set, several such, one or several bathrooms per house, a cat, a dog, are legitimate wants. Decisions about targets for the amounts of resources to be used would necessarily be arbitrary, bureaucratic ones, related to the level of resource use that the bureaucrats arbitrarily deemed acceptable. The arbitrariness springs from the limitations of our knowledge. If we were omniscient and all-wise, we should know what was the right amount of saving of resources for the future. We and the bureaucrats of the resource-frugal society are not omniscient. The latter could be seriously in error in their judgments, yet we could have no rational basis for challenging their decisions. If they opted to reduce waste as much as possible, if they set very rigourous, austere standards, with production limited to meeting minimal demands, we could have no rational grounds of appeal. Resource-frugal bureaucrats would view with disfavor, as being unnecessarily wasteful, interesting, pleasure-causing changes in fashions and tastes. They would have to accommodate production to take note of varying weather patterns and the demands they create, yet they would necessarily not be able to permit in their system the flexibility necessary to respond fully to the very considerable climatic variations that occur from year to year.

Such a centrally planned economy, which controlled resource use and to a degree restricted growth, and directed production to meet predetermined targets based on the planners' decisions about needs and legitimate wants, would necessarily be one that curtailed freedom—the freedom of producers, farmers, manufacturers, even inventors, the freedom of the consumer, the freedom of the ordinary citizen in exercising property rights in money, land, house, investment. Persons would be restricted in the energy they could use, in the energy-using activities in which they could engage. They would have vastly less choice than they now have in respect of the goods they could buy, food, clothing, furniture, appliances, cars, houses. There would obviously be restrictions on the building of new houses, and greater insistence on reusing the old. If such a society were not to be even more inegalitarian than are our present societies, it would need to adopt a rationing system for consumer goods of the kind that prevailed during World War II in Britain. Otherwise, the poor could be outbid for most goods by the wealthy. The loss of liberty would be a loss of both freedom from interference and positive liberty to be effectively self-determining. The bureaucrats, not the individual person, would be the master of his destiny. As they gained confidence, and the ordinary person came more and more to lose his capacity for initiative and enterprise, the bureaucratic planners would come to run the resource-frugal society as factory-farm managers with economic frugality manage their factory farms, shaping the needs and wants of their charges to meet the economic requirements of factory production. Man of the resource-frugal society could come to be preplanned, predetermined, in a way comparable with that employed by the planners of factory farms in respect to their animals.

It is worth noting that if the facts concerning the imminence of a resources crisis were as the crisis theorists claim them to be, then a Millian liberal, in terms of the Millian prevention-of-harm principle, could not seriously object to the gross, far-reaching invasions of liberty that are proposed as necessary to prevent the disruption and harm that would come from the exhaustion of key nonrenewable resources. This fact alone must cast very serious doubt on the soundness of Mill's philosophy of the state as a basic statement of liberalism.

Various of the important costs that arise from implementing a policy of conservation of resources relate to the need to have, and the problems involved in securing, the adoption of this policy internationally, and the costs associated with achieving this. However, before looking at these problems and their costs, it would be convenient to review the costs of resource-frugality at the national level.

The great cost in terms of loss of freedom, positive freedom to be rationally directive and master of our destinies to the degree we now are, and of negative freedom, to be free of undue interference, has already been noted. An aspect of this loss of freedom, which has already been alluded to, needs more explicit noting, however. The great, necessarily arbitrary, power placed in the hands of political leaders and their bureaucrats in a resource-frugal society constitutes a serious, dangerous loss of freedom and an objectionable source of danger to individuality. Few persons with any political awareness would be willing to entrust to a state bureaucracy, even less to a world community of nations bureaucracy, such important, immensely consequential decisions. This distrust is well founded, based on solid experience of bureaucracies and bureaucratic mentality. Bureaucrats would, increasingly, simply impose their preferences on the community; they would increasingly disregard the consumer's needs, wants, rights. Innovations would increasingly come to be seen as undesirable nuisances that complicated planning, as something to be resisted, discouraged, ignored. Democracies have as yet to find a way of effectively making the bureaucrat serve the community first and his bureaucratic class interest last.

The problems in respect of unemployment, and the costs and difficulties in the way of ensuring that all persons in a resource-frugal society can live with dignity and self-respect and without fear of destitution and all the evils associated therewith, would be great.

The problems concerning possible unemployment and adequate welfare benefits relate to equality and justice. However, the resource-frugal society gives rise to equally basic concerns about justice and equality. A resource-frugal society would be an economically limited-growth society; it would admit of some growth, but much less than a resource-liberal society. There are two serious disadvantages from the point of view of justice and equality of such a limited-growth society. The one is that it makes it more difficult for there to be mobility of individuals between economic classes and for there to be changing economic relativities of groups of persons. With significant growth there is a great deal of individual and group mobility upward and downward. This has great value as permitting justice to come to those who merit it. And it has great value as helping to lessen social friction and discontent. Where it is known that ability, industry, and enterprise can find their own level, there is greater social acceptance of the economic order, and greater social harmony. In growth economic societies, there is a good deal of movement of this kind, involving groups as well as individuals. This prevents or lessens the ossification of societies into caste-type classes.

The second limitation of a limited-growth society is that it makes changes of relativities of the kind dictated by justice, and a general lessening of economic, social, political, and status inequalities of the kind dictated by justice, more difficult. Growth allows for more change, more flexibility. This is in no way falsified by the fact that inequalities have persisted in growth societies. The families and classes involved in the equalities have to a degree changed, but glaring, unjust inequalities have continued. In Europe and North America (apart from Mexico) there is not the destitution that prevailed in the nineteenth century. And the development of the United States and most states of Europe as welfare states has meant a levelling up of some importance. However, whatever is true of the past, it does remain true that there is a greater scope for peaceful, lawful reduction of unjust inequalities in societies in which there is growth. There the inequalities can be reduced without taking wealth away from those who possess it. Many who seek to underplay the capacity of growth societies to lessen unjust inequalities write as if the problem of inequality in growth societies is as it has always been. This is not so. There are new, different inequalities, affecting different groups in society. Our growth societies are dynamic, fluid societies, societies that allow mobility, adjustment to correct injustices, and real opportunities for enterprise, initiative, and ability.

How is a resource-frugal society to be realized? Here we need to consider how the problem arises and might be dealt with at a national level, for example, in respect of the United States, and then at the level that really matters if conservation of resources is as important as its exponents claim, at the world level. Serious discussions of how a resource-frugal economy is to be brought into being at a national level are notable by their absence. As M. E. Kraft in "Political Change and the Sustainable Society" observes, what we find of writers who have the U.S. most evidently in mind is expressions of hope, vague recommendations, and the like. To this end, Kraft noted:

Most of the authors cited above [R. O. Loveridge, W. A. Rosenbaum, M. J. Brenner, L. K. Calwell, D. H. Henning, W. Ophuls, D. C. Pirages, P. R. Ehrlich, M. Mesarovic, and E. Pestel] conclude their discussions with a set of recommendations which they predict (or simply hope) will alter the unfavorable climate of present politics. For example, we are advised that we need more central coordination of policy making and possibly national planning agencies, or even extremely powerful new forms of government—largely authoritarian in nature; that we need to reform Congress and the "captive" regulatory and administrative agencies to "open up" or democratize U.S. government; that we need to turn from excessive dedication to individualism and laissez-faire economics to more cooperative social values and socialist economies; that we need to arouse and educate the U.S. public and create a new environmental consciousness and a new environmental ethic—borrowing heavily from Eastern philosophical and religious traditions to challenge Western rationalized, materialistic cultures. [Pirages, 1977, 186]

Kraft's concern relates not simply to getting the desired measures implemented but also to making sure that the desired measures will bring the benefits that are sought from them. Clearly, given the costs inherent in the adoption of a resource-frugal economy—possible population controls, severe restrictions on freedom, greater unemployment and dependence on welfare benefits, less mobility and scope for lessening unjust inequalities—it is highly likely that any political party that went to the electorate with such a program as its proposed policy would be destroyed politically. That is why one so often encounters hints of a resort to undemocratic, authoritarian, totalitarian solutions. There are, however, no easy ways of bringing about a political revolution based on a concern for conservation. Further, the costs involved in abandoning democracy are immense and could in no way be justified in terms of the goods to be achieved by a policy of conservation. And dictators are notoriously uncontrollable by their supporters, and they are much given to going their own ways, whatever may have been the intention of those who brought them to power.

Equally little, or perhaps less even, is said about the problem of bringing the world community of nations to adopt a policy of conservation of resources. That the problem there is no small one is evident from the very limited success interested parties have had in securing cooperative world action to conserve even such a minor resource as the whale, and the even lesser success in conserving adequate numbers of fish species. What hope of success then is there of joint world action for the sake of such basic resources as petroleum, coal, mercury, zinc, and the like? Some of those few who attend at all to this problem seem to think that if the United States and perhaps some other resource-profligate nations could be persuaded to opt for a resource-frugal economy, it might be possible for the U.S. to coerce other states, more evidently those in the Third World, into curtailing their economic growth and their use of nonrenewable resources. There are obvious moral, political, and practical objections to the idea of one or more nations so coercing other nations into adopting a resource-frugal economy. The nations of the Third World—India, Brazil, Mexico, Pakistan, and the like—would not and morally could not freely agree to opt for a resource-frugal economy, given that increasing use of resources and economic growth are vital if their citizens are to have any chance of enjoying their human rights. Exponents of the no-growth doctrine show an incredible insensitivity to the needs and rights of such nations and their citizens. It is not only the nations of the Third World that would need to be coerced.

Coercion would have to be applied against resource-profligate nations that failed to mend their ways; it would also have to be directed against the USSR and nations of the Soviet bloc as well as against China, if they did not voluntarily cooperate. Economic coercion would not be enough. Only war and the threat of war would be adequate. For that to be adequate the so-called committed resource-frugal nations would have to commit immense amounts of resources to preparedness for war and to the conduct of war. Moral objections to the use of such coercion relate to the lack of right of any nation to impose its will on another nation that is not an unjust agressor and is not unjustly oppressing its peoples. They relate also to the very great evils of wars, of death and injury to persons, and serious, lasting damage to the environment. If the concern for conservation of resources is to be a serious world concern, one that takes full moral note of the rights of persons in the Third World, better-thought-out, morally more serious proposals than these are needed. It is hard to see what measures that are morally and politically acceptable could succeed in gaining the necessary world acceptance.

The case for seeking to opt for resource frugality on a national and on a world scale has been argued against on the bases of practical, moral, and political objections, that on all these counts the price to be paid is far too high for the benefits sought, and even more, for the benefits likely to be gained.

This may be interpreted as an indirect endorsement of the resource profligacy that now prevails in the Western world. That is not so. What is being argued for is a resource-liberal, not a resource-profligate, society. Much could be done by way of reducing waste without reducing standards or levels of enjoyment. Instead of incentives to waste, there could be introduced *dis*incentives designed to discourage waste. Charges for energy could be made to increase per unit with the increased amount used. There could be tax disincentives of many kinds to encourage the use of less powerful, lower-gasoline-consumption cars. Built-in obsolescence can be legally banned. Incentives to foster recycling of materials could be greatly increased, and much done to have such recycling itself powered by renewable energy resources. Much could be done by way of government incentives and regulation to bring about greater use of solar energy, as in heating buildings.

Building regulations could come to include requirements about insulation and design that reduced heating and cooling energy demands. And so on. The switch from a resource-profligate society to a resource-liberal one would involve an acceptance of much more state action, more state interference by way of incentives, disincentives, regulations, restrictive laws, and overall planning than now prevails. However, it would not require the loss of basic 'freedom, nor the acceptance of high levels of unemployment, nor the closing or lessening of opportunities for social mobility and removing unjust inequalities. And it would leave us free to accept our responsibilities toward the Third World. We would in no way seek to impede those nations in their efforts also to become prosperous, resource-liberal societies.

12

Science and Technology: Political Controls?

As noted in Part II, many environmentalists, although greatly impressed by the science of ecology and the value of the knowledge that ecology gives us, express a fear of science and technology, a fear that science may give us knowledge, and technology powers, that we cannot handle. Further, there is a fear that we may come to be controlled by our technology. Various of these fears, both about the knowledge that science brings and the powers that technology confers, are not without foundation. Genetics and nuclear physics have provided knowledge that has far-reaching consequences and implications, while many other sciences have provided knowledge of far-reaching importance. Technological discoveries have greatly changed society and no doubt will have great effects on man's future mode of life. Technological developments can bring about the need for very rapid social, economic, and political changes, such that our responses may be inappropriate or too slow to avoid serious hardship for many. This seems already to be the case in the resistance being offered throughout the industrial world to moves for the shorter working week that technology makes possible and morally, socially, and politically desirable.

It needs also to be acknowledged that social and other changes that follow new technologies may be those best suited to the operation of the technology rather than those most beneficial to man. What occurred in the industrial revolution will no doubt again happen with the technological revolution of automation, computerization, microcomputerization. The important message to be gained from past experience is not that we should seek to resist technological advances—that is both futile and self-defeating as well as denying many persons, for a time at least, many important goods, not least of which is leisure of the kind that Aristotle rightly saw to be essential for human well-being and the good life. Rather, we should seek to recognize what science and technology can do for man and use them to further the interests and well-being of all persons. This will mean making the appropriate economic, social, political, and personal adjustments. Arrangements that worked well in an agricultural society do not necessarily work well in an industrial society. What worked well in a precomputer society is unlikely to work well without modification in a computerized society. The technological revolution that can be aided or resisted by the state can bring great benefits in terms of improvements in agriculture and food production and in the production of energy and goods, especially in bringing into use energy from natural, nonpolluting phenomena such as the sun, winds, tides, and geothermal sources. To look more closely at the appropriate role of the state in respect of science and technology.

Science

In spite of their great faith in the value of the knowledge that the science of ecology brings, many ecological moralists distrust science in general very deeply and would have the state check and control research, especially in the frontier areas of nuclear physics and genetics. This distrust of and desire to curtail scientific research is dangerous and morally indefensible and is reminiscent of the reaction of the Church to science in the time of Galileo. Man should have no fear of new knowledge, but he should have a fear of placing control over knowledge in the state. The state lacks the competence and the right to determine what new knowledge we have no right to possess as being too dangerous for us to acquire. Further, among the basic rights of man is the right to access to knowledge. We need to gain as much scientific knowledge as possible to manage our lives and our living in the world well. It is not improbable that ultimately, through the investigations of the sciences, we will come to know how to reproduce in the laboratory processes such as photosynthesis and synthetizing fossil fuels, to have a full understanding of the earth's weather and climatic patterns, and the like. Man will be able to use this knowledge for his and the earth's best advantage. Consider how reliable long- and short-term weather forecasting could lead to improvements in agricultural production. Such knowledge, like all knowledge, will carry with it dangers of misuse. An understanding of weather and climatic patterns may well give man the power to create droughts or floods in countries of his choice, to initiate an ice age or a calamitous rise in world temperatures. However, this fact constitutes no reason for holding back the pursuit of knowledge, for entrusting rulers with the control of science.

Technology

Technology and the powers it confers on man are similarly distrusted. Here there is a case for action by the state to ensure that the discoveries of technology are used for man's best advantage. There are real dangers that new technologies will polarize wealth and power on the one hand and poverty and political impotence on the other, as they did for a time in the industrial revolution. Yet modern technologies can be made to work to aid man's self-development. They can release persons from much mindless, degrading work. Hence some, even many persons may never need to work. Their self-development can come in other ways, without lack of respect or self-respect. What it is important for the state to ensure is that work is shared, that the benefits of technology are justly shared, and that the leisure that technology makes possible be seen not to place any category as "unemployed" in the present sense, but as "leisured," ensuring that all have status and dignity as persons. It will therefore be necessary for states to train and equip their members for leisure, such that it is used in meaningful ways and not simply for spectator amusements.

Up to the present time, the democratic states of the world have, except in matters relating to defense and the support of universities, largely left the development of technologies to private enterprise. They have allowed private enterprise to conceal, and even to buy out in order to render unavailable, useful new technologies; and they have permitted it, by way of patents, to cost valuable technologies out of the reach of many. A well-ordered state will foster the development, testing, and application of useful technologies, particularly but not only those in the important areas of energy production and the locating and extracting of resources. This means that the state will need to be active and interventionist in the sphere of technology, acting to ensure that new technologies be developed and used and their benefits widely enjoyed. States can do much more than they now do to foster the development of new technologies. This is evident from what they achieve in technological research in war and in peacetime defense research. Private enterprise cannot safely be entrusted with the major responsibility for the development of technology. It pursues technologies for private profit, not for the common good. This leads to the neglect to follow up, even to suppress, useful but for them unprofitable, technologies, and it leads to a concern for quick rather than long-term developments on the part of many corporations.

An increasingly important area of technology relates to eugenics and to genetic engineering. Eugenics has already been used to develop technologies to increase food production in innumerable ways, not the least being the producing of the many new varieties of plants and animals, including those that made possible the Green

Revolution. Much more no doubt will be achieved by scientific breeding of new varieties of plants and animals. Genetic engineering, on the other hand, constitutes a dramatic, exciting breakthrough of major proportions, one that opens up vastly wider possibilities for developing new plant and animal organisms. At this time, it is not clearly known what is the full range of possibilities opened up by genetic engineering, other than that it may well revolutionize medicine, food production, industrial technology, and energy production. What its implications are in respect of the development of man from *Homo sapiens* to *Homo geneticae* is entirely a matter of speculation.

Because of the dangers to mankind of much research by way of genetic engineering, and because of the immensity of the powers it appears likely to confer on mankind, responsible states are already rightly claiming the right to supervise and insist on safeguards in respect of such research. Issues also arise here concerning the moral and political desirability of allowing property rights by way of patent rights to discoveries. Decisions have already been taken. They cannot reasonably be objected to without the whole private-property system of liberal democratic states being questioned.

The application of genetic engineering to human beings appears to lie in the less immediate future. Nonetheless, it may come to be possible to make drastic changes to human beings. If this should prove to be the case, the new powers will be very consequential, raising basic moral and political questions. Genetic engineering may well create a human drone. If so, states would be faced with the issue of whether to allow such drones to be multiplied and used in industry as industrial slaves or the equivalent thereof. States will also be faced with the ecological impact of many products of genetic engineering. At present there is awareness of the dangers to man inherent in present research. There will need to come an awareness of the indirect dangers to man that come with threats to his environment and to the organisms that now inhabit it, and of the need for relevant controls and checks.

If the ecological state were to be as totalitarian as many of its exponents suggest in its range and exercise of controls by means of very pervasive, invasive restrictions on individual liberties, autonomy, privacy, and the like, it might be wondered whether such a state might not go further, and foster, even enforce when that comes to be possible, the use of genetic engineering. Such interference would not be out of harmony with the massive controls that many environmentalists already insist are necessary. Relevant points here include: (1) There is no logical necessity, in terms of consistency, for the ecological state to take such a step. On the other hand, it would not be an unreasonable one for it to take, when its major concern is represented as that of achieving an efficient, well-run economic community at harmony with its environment. Utilitarian ecological moralists may well favor such a concern by the state as dictated by the maximizing of pleasure. (2) If this approach is thought to be desirable, we need not wait for advances in genetic engineering to occur. We can already encourage the state to use the available knowledge of eugenics to weed out from the human gene pool such genetic weaknesses as diabetes, hemophilia, and the other hereditary diseases. If the state is to plan society efficiently, its rate of growth and non-growth, its population, its actual production of goods, and the like, it would seem reasonable for it to go on to plan the production of healthy, genetically sound human beings, and superior human beings, or human-type beings if genetic engineering makes that to be possible. Coercive methods of achieving these ends would no doubt be necessary. They would involve vast encroachments on human rights and human liberties. They are to be rejected on that count. That there would be grave dangers in entrusting such powers to states is a further reason for not doing so.

Technology and Factory Farming

Arguments have been advanced against factory farming and the state permitting factory farming, it being urged that factory farming should be banned by states on the ground that it is wasteful of food due to the food lost converting grain and cereals to meat, and also on the distinct ground that it involves great, avoidable, and unjustified cruelty. Both arguments against factory farming are contingent ones dependent on the continuing truth of factual claims. At present, factory farming is an economical way of producing meat, but it is wasteful of food in the process. However, if the development of microorganisms that permit the use of formerly useless vegetable matter such as stubble to produce microorganisms that in turn can be used as food in factory farms succeeds as now seems very likely, factory farms could come to produce valuable, needed food from non-food. In a world in which it is widely claimed that there is not an abundance of food, it would be wrong not to use such a means of producing food. At present, factory farming is at a very early stage of its development. Much more research and experimentation needs to be done to determine which of the various animals suitable for factory farming can produce the most meat most economically. Rabbits are claimed to be efficient converters of vegetable matter into meat; if this is so, they might well be used in factory farms. Clearly, many avenues of possible development need to be explored.

The second argument against factory farming stresses the cruelty inherent in it. States are rightly seen to have a duty to prevent avoidable cruelty. Against this argument, it needs to be acknowledged that while much factory farming is at present cruel, not all of it is such. Further, it need not be cruel, certainly not as cruel as it commonly is. No doubt lessening and, even more, eliminating, the cruelty would lead to significant increases in the cost of factory-produced meat. However, as has been urged by R. G. Frey and others, there may be or come to be another way of reducing or eliminating the cruelty, that of developing animals, not necessarily fowls, which find the economically cheapest factory farm environment a congenial one. The "invention" and designing of such animals seems now to be feasible. If such are produced and used, this moral objection to factory farming would cease to have any relevance against this kind of factory farming.

Having noted these considerations, it does need to be stressed that one role of the state consists in the preventing of avoidable evils. Much avoidable evil in the form of cruelty in the production of meat, farm produced and factory produced, occurs that ought not to occur. States have the right and duty to check this cruelty. The reduction of such suffering must increase the costs of producing meat and thereby make meat less readily available to those who most need it, forcing them to rely on a predominantly vegetarian diet. This, in turn, would have the ecological consequences noted earlier.

The pollution aspects of factory farming are not unimportant, since the use of factory farming contributes in three different ways to pollution not to be found with free-range farming. Factory farm animals feed on cereal food produced using pesticides and fertilizers, both of which may cause pollution. Their excrement constitutes another serious form of pollution. In a well-managed economy, that, far from being a source of pollution, would be used as a resource.

13

Pollution and Pollution Controls

No new political issues of principle are raised by the problem of pollution other than that some of the measures that are necessary for effective pollution control involve restrictions on freedom of kinds traditionally minded liberals accept only reluctantly. Rather, the political issues of importance arise from the variety and forms of pollution, and from the need for a very complex, multi-level, multi-pronged approach. Pollutants range from very dangerous poisons that threaten health and life and create conditions that make impossible the full enjoyment of persons' other basic rights to pollutants that simply constitute a nuisance that takes away from the quality and enjoyment of life. Pollutants may be widely dispersed in the atmosphere and in the oceans. Or they may be more or less relatively localized to a geographical area, to a city, even simply to the environs of a factory or energy plant. They may be enduring, lasting long after the activity that caused them ceases, or they may quickly be dissipated once the activity that produces them ceases. The harmful effects of pollution may endure long after the pollution has ceased to occur, or they may lessen and soon no longer occur once the pollution is no longer produced. Pollutants may be directly harmful to persons and their property, or only indirectly so, by way of harming nature and the natural environment. Pollution may harm only presently living persons, or it may also or even only threaten the well-being of future generations. And the pollution that is produced may be quite unnecessarily produced, being such as to be easily and economically preventable, or it may be an unavoidable/difficult-to-avoid accompaniment or effect of desirable activities such as food and energy production.

This great variety in the range of forms, kinds, of pollutants makes a multi-level, complex approach to the problem of pollution politically essential. Much pollution calls for an approach at an international level, by cooperative action between all nations. Some simply requires cooperative action by those nations affected by the pollution. Much pollution is best to be combatted at the national level by the nation-state. However, with other, localized, pollution, often the local authority is well placed to cope with the problem. Complicating the issue is the fact that states are themselves great polluters, and both as individual entities and collectively often act with great irresponsibility, as in using the atmosphere and the oceans as sinks in which to dump highly dangerous, long-enduring pollutants. Effective pollution controls will ultimately involve controls on the state, a world antipollution authority with real powers to enforce conformity with its directives.

There is some dispute as to whether man has the technology necessary to control and check pollution. The weight of evidence would seem to be that man either has the technological know-how to lessen pollution to acceptable levels or where this is not so can, at some cost, dispense with those activities that cause the pollution. The real problem is the political one. In democratic states that problem is the problem of gaining acceptance for the measures needed. Since the general public, with their demands for energy, use of cars, detergents, pesticides, demands for consumer goods produced by polluting processes, are major contributors to the production of pollution, and since the reduction of pollution will increase costs of production and hence be resisted by producers among whom may be the state, there are problems in having the costs of pollution accepted. That the corporations that most offend in this matter are often politically powerful, and that the state itself is so commonly a major offender, make the practical political problem a major one. Many despair of adequate solutions being achieved through the democratic process. Yet, when it is appreciated how recent is the escalation of pollution and of the general development of a pollution-consciousness, the progress made in developing antipollution measures, while well short of being satisfactory, is not unimpressive. Prevention of pollution has become what it must be, a serious political issue.

States have already developed instruments with which to check pollution. What is needed is a more concerted, more consistent, more persistent, more thorough use of the methods already devised, with the penalties attached to the criminal sanctions being increased to be commensurate with the nature of the crime, where the crime of pollution is the crime of killing, injuring, maiming persons, taking away from the quality of their lives. The range of approaches already used includes the dissemination of information and the educating of the public and other polluters alike. Incentives and disincentives of various kinds are also important. The state can contribute to the reduction of pollution by providing aids of various kinds, including subsidies to those who cannot bear the costs of ceasing to cause pollution and whose activities have social value. However, subsidies are not the only form of aid that is needed. If pollution from pesticides is to be reduced, a great deal of research of many kinds must be fostered: research into biological controls, research into pest-resistant plants, and the like, research of a kind in which the farmer who causes the pollution cannot himself engage.

Although the criminal law is a clumsy instrument, and one that is unsuited to be the only weapon to be used against pollution—its use can result in gross injustices and in unacceptable costs—it must nonetheless play a major role in combatting pollution. The causing of pollution, and not only dangerous pollution, can rightly be made a criminal offense as an act of harming others and violating their human rights. And such offenses can and ought to carry with them the same penalties and the same legal stigma as do comparable criminal offenses of killing and harming others, or of causing serious nuisance. However, even where it is just and appropriate, this direct application of the criminal law may not be adequate. Much serious pollution is the coming together of small, seemingly harmless amounts of pollution. While there can be no moral objection to making it a crime for car owners to have and to use cars that cause pollution, it is more effective to have, in addition or instead, criminal laws that ban the manufacture and sale of cars not fitted with antipollution devices, and also to ban the manufacture and sale of pollutant leaded gasoline and the like. So too it is often vitally important to prevent the pollution occurring at all. Since the deterrent effect of punishing polluters is only of limited success, a ban on the activity may be necessary. Further, the potential polluter may be undeterred by criminal sanctions because of overconfidence about his ability to prevent pollution occurring. (Such a use of bans can be defended in respect of the building of nuclear energy plants, with their dangers and dangerous wastes, more especially where alternative fuels are available.) This approach involves serious restrictions on liberty of a kind liberals already accept in other areas, but often, although not always, with some reservations. (Consider the divergent attitudes of American and British liberals toward gun laws, where the same sorts of issues arise.)

Criminal laws directed at punishing those who, by creating pollution, harm others need to be supplemented by laws making it easier for those harmed to gain reparation for harm done. This will not always be easy or even possible. Often the harm done is slight, but such as adversely to affect many thousands of persons and their property, where the total cost to others is great, but to each individual relatively small. It is here that class actions with parties affected being given standing could play an important role, both in securing justice and in deterring offenders.

All these measures need to be reinforced by a strong, independent, well-financed antipollution agency, the role of which is to inform itself about pollution and developments in that area, to monitor pollution levels, to issue warnings when it foresees problems in the future, to recommend those antipollution measures it deems to be necessary, and to initiate prosecutions against those who offend against antipollution laws, it having the power to act against the state and against state officials. Such an agency would have to be a state instrumentality, ultimately responsible to the people through the state. Yet, if it were to be effective in checking pollution, it must have considerable independence and autonomy to act against the state itself and those corporations who so often have powerful friends in the legislature and bureaucracy. Democracies are taking timid first steps in that direction, setting up weak, toothless agencies. However, given the power of bureaucracies to enlarge and strengthen themselves once established, these modest agencies may grow into the powerful organs of the state they ought to be, the more so if they gain the public support they need.

Ultimately, whether there come to be adequate controls on pollution will depend on public awareness, public interest, public concern. For this reason, although its immediate impact may be small, ultimately much will depend on the public being fully and adequately informed and educated about the facts concerning pollution. Here as elsewhere, misinformed, alarmist claims will do the antipollution cause no good. The truth is what is needed, and all that is needed, provided it is widely disseminated.

Like the proposals relating to securing preservation for man's well-being, measures to bring about a change from a resource-profligate to a resource-liberal society, and controls to ensure the best use of technology, the controls advocated to reduce pollution constitute an acceptance of further checks on freedom, greater centralized, governmental control, such that the ecologically informed, liberal, democratic state is vastly different from the modest state conceived of in traditional liberal theory.

14

Population, Overpopulation, and Population Control

In Part II it was argued that a tenable, plausible case had not been made out for believing that the world is threatened by a population crisis in the foreseeable future. However, it was acknowledged that mankind has the capacity to overpopulate the world. For this reason, and because the constantly recurring and reiterated predictions of a world population crisis have led to a very considerable literature, indeed, a veritable industry, devoted to setting out the moral and political "reforms" that this crisis makes to be necessary, where this literature attracts a good deal of public attention, and yet where much of it is ill-conceived, ill-considered, and mischievous in its effects, it is desirable that serious consideration be given to the nature of the measures that would be morally and politically legitimate and desirable, and politically feasible, should the world be confronted by a population crisis. That there may be an independent case for aiming at an optimum world population also makes to be morally worthwhile an inquiry into measures to control population growth.

In approaching a consideration of these issues, it will be useful to recall various of the points developed in the discussion in Part II. There it was argued that the person has a right and duty to produce only that number of children for whom he or she can adequately care or for whom he or she knows there will be adequate care. However, this is qualified by the rights of all persons to a just share of the world's resources, in particular, to what is necessary for a decent, self-developing life (unless, by their actions, they have forfeited this right). Persons have a right to respect as persons, to personal autonomy, and to the liberty and opportunity to be self-developing. The desire to reproduce, to have a family, is a basic, rational, human desire, and such that action on the basis of it relates in an important way to a person's autonomy. For many, the family is an important context of self-development. Hence, as a matter of respect for human rights, and as a matter of justice, all should without arbitrary discrimination enjoy the right to have offspring. While the right is qualified by the duty, there is the further qualification that the child, once born, whether or not rightly brought into being, has rights, not least rights to life, health, and respect as a person. The right to reproduce, although not itself *sui generis* a basic right but an aspect of basic rights, is like them a prima facie right which admits of being overridden by other moral rights and other moral considerations. However, to be morally overriding, they need to be powerful moral considerations.

This position is distinct from that of J. S. Mill, who appears to have seen the right to marry and to reproduce as simply an aspect of the right to liberty, and hence as one that was qualified by the harm principle, the right failing when its exercise caused avoidable harm. Hence it was that Mill proposed a means test—if not inexpedient on other grounds—for those who contemplated marriage, seeing the application of such a test as preventing "the mischievous act" of bringing into being children for whom their parents could not care. (See Mill, 1848–71, II, 12, 2.)

The thesis to be developed here is that if there were to be a world population crisis, voluntary methods of persons freely choosing to limit births by sexual abstinence, natural and artificial methods of contraception, sterilization, and abortion, even if reinforced by education and making contraception, abortion, and sterilization freely available, will not succeed, even if backed by a system of incentives and disincentives, in terms of taxation inducements, housing, job opportunities, and the like; where such incentives and disincentives may well operate to the detriment of innocent children; and where, if they worked, they may well work by encouraging persons to opt for high-resource-use luxuries rather than for children, and hence thereby thwart one of the aims of population control. That voluntary methods of checking birth rates are unlikely to be adequate is suggested by the relatively limited success of family planning projects even where they are backed by information, education, and making available contraception, abortion, and sterilization. Family planning and its success or lack of it throws light on the likely success of voluntary population control programs, even though family planning is something quite distinct from population control, indeed, being the antithesis of it. Family planning rests on the view that the individual family has the right to determine the size of the family, be it a family of one or twelve, and to determine when each child will be born. It therefore has the high motivation of being dictated by the wishes and interests of the family that makes the decisions.

With population control to prevent a population crisis, the decision as to how many children a couple may have is determined not by them but by state officials on the basis of their estimate of what is compatible with ultimately stabilizing the world's population at what they judge to be the right level. Individuals are called upon to align their decisions and their wills with what is judged objectively to be necessary for the ultimate goal of a stable population. The realization of this goal may dictate fewer children than the persons concerned wish to have. Thus their motivation in aiming at the number prescribed by the state, if there is no general coercion, will be a sense of duty and public spiritedness, not self-interest or motivation from self-chosen goals. The former motivations are likely to be much weaker than the latter. It is, of course, possible that if there really were clear evidence of a danger of overpopulation this might strengthen the former motives and bring the latter more into line with them. However, the facts of life and birth rates of seemingly overpopulated countries today give little reason for believing that this would be so.

Even when highly motivated, family planning is much less than one hundred per cent successful. Many who practice family planning end up with larger families than they intended to have. Many more who planned to determine the size of their families using contraception end up gaining their goals or a family of one more than their goal by resorting to abortion, and less commonly to sterilization. Abortion is a key part of family planning in the countries of the Soviet bloc and in Japan; it is a vital, much used, back-up measure for family planners in Western democracies such as the United States, United Kingdom, Australia, and New Zealand, being an important element in the family planning of the unmarried. Birth rates would be significantly higher in the latter countries if abortion were to be outlawed effectively. Thus, even if all persons were to be persuaded to practice family planning, that would not provide grounds for believing that voluntary methods of checking population growth would work. Family planning and population control have different goals as well as different motivations, such that the world population that would result from the universal, successful practice of family planning would very likely be much larger than that desired by those who claim there is a population crisis. It is unlikely that a resort to abortion or sterilization would be as acceptable to the persons concerned for purposes of population control as distinct from family planning.

The fact that hundreds of millions of persons do not practice family planning is also relevant to the likely success of voluntary methods of population control. In particular, it has not caught on to a high degree in high-population-growth countries, and this in spite of campaigns to achieve this. Available information suggests that in many cases, where it is accepted, it is only after a large number of children have already been born.

Even if states were to engage in high-pressure propaganda, even something approaching indoctrination, as a means of instilling a sense of social responsibility about population growth, so that persons saw it as a matter of concern to society and not simply to themselves as individual persons how many children they had, it is unlikely that voluntary methods of birth control would succeed as methods of population control. In any case, there are evident dangers in, and moral and political objections to, the state resorting to indoctrination or anything approximating it, no matter how worthy the end toward which it is directed.

Much is made by some writers of the possibility of and likely success of a scheme of incentives and disincentives which may approach being coercive in their operation. The kinds of things that are mentioned here are restrictions on housing for large families, tax disincentives, withholding maternity and welfare benefits for "excessive children," lessened job opportunities, a child tax, high-cost marriage licences, and the like. Besides having the two disadvantages already noted, if they work they may drive expenditure into high-resource-use alternatives to children, and if they fail effectively to work they may harm innocent children; the available evidence is not encouraging about their efficacy.

The confidence some writers have in such methods links with the much more widespread, dangerous, ill-based faith so many who believe that there will be a population crisis have in the capacity of governments to influence birth rates significantly and easily. It is true that society, social attitudes, which can be influenced over a period of time by the law and the state may influence the attitudes and behavior of individuals regarding reproduction, but states—as distinct from societies and cultures—in their deliberate attempts to increase or decrease birth rates have generally had little influence other than when responding to, as distinct from forming, popular demands by way of incentives, disincentives, and restricting or relaxing access to contraception, abortion, and sterilization. China is an important exception, but the price paid there is very high both morally and politically. Japan might seem to be another notable case of a country in which government policy was instrumental in reducing birth rates. In fact, the lower birth rate is the result a morally dismayingly heavy use of legalized abortion by persons engaged in family planning, where the legalizing of abortion was itself not a part of a government plan to foster population control but a response to popular demand for easy abortion for family planning.

The arguments to date have related to the inadequacy of voluntary methods of population control, if there were to be a population crisis. Those who believe that such a crisis is imminent generally would not dissent from this claim, although their reasons are rather different. They stress the urgency of the need for decisive action, the desperateness of the situation, the time-lag that voluntary methods involve before they begin to operate at all effectively. It is therefore to the methods that rest on the use of coercion that we need now to turn our attention.

Most of the methods that involve coercion are morally unavailable, and are such that the state can have no moral right to employ them. They are also such that it is politically completely unrealistic to expect them ever to be politically acceptable or feasible. The only coercion that morally is open to the state in the area of population control is that involved in making it a crime to have children in excess of the permitted number where that is rationally determined on the basis of good evidence of the reality of the danger of overpopulation. This means that it would be up to the individual persons to determine how they conform with the requirements of the law. If the unjust, antiegalitarian consequences likely to follow them allowing the practice could be prevented, such a law could be associated with a system of transferable baby licenses. This approach could also be linked with a system of incentives for late marriage and the like. In one respect, such a law would be easier than most laws to police, in that offenders (or at least one of them) against it would be easy to detect. The two problems that would arise in respect of this approach are, first, that the law and law enforcement could carry the burden such a law would impose on them only if it were seen evidently to be the case that the law was necessary to avoid the evils of overpopulation, and, second, that it could lead to much hardship and suffering, although much less than that which genuine overpopulation would cause.

The other aspect of this contention is the vast range of bizarre, morally outrageous proposals that are flaunted in the literature concerning the population crisis, that there be compulsion in the use of contraception, that young women be temporarily sterilized against their wills, that those who have their quota of offspring (or one in excess) be sterilized, that there be sterilants in the drinking water, that abortion be forced upon those who otherwise would exceed their quota of children, and the like. The interesting, difficult moral problem is that which would arise in determining the nature of man's moral rights and duties in respect of those countries that continued to permit overbreeding, overpopulating in a world that was overpopulated or on the brink of overpopulation. Many writers already contend that today states with surplus food and medical supplies have the right and duty to withhold aid until states that are overproducing children come to adopt effective population measures. There are serious objections to such a thesis, but it at least has the merit of facing what in the event of world overpopulation could be a major moral issue. Before this issue can be pursued, the case for rejecting the coercive methods noted above needs to be filled out.

Compulsory contraception, sterilization, and abortion are morally unavailable because they would involve violation of basic and derivative human rights, rights to respect as persons, to bodily integrity, to privacy. Unless the state were prepared to violate personal privacy extensively in morally quite unacceptable ways, legally to enforce the use of contraception would amount to punishing those who had more than the permitted number of children. It would involve the injustice of punishing the very many persons who conformed with the law and used contraception, where the contraception failed. The enforcement of the use of artificial methods of contraception is also morally and politically objectionable, as involving the forcing of the conscience of such persons as strict Roman Catholics. Were no other course open to the state, and were it more efficacious than any other, consideration could be given to whether the goal of a stabilized world population justified the forcing of the consciences of hundreds of millions of persons, as well as the other evils this approach would involve. That the effective use of artificial methods of contraception requires a high degree of motivation and some degree of understanding of its mode of operation makes it clear that even if it were morally permissible to impose it on unwilling, uncooperative persons, it would clearly fail in such countries as Mexico and Brazil, as well as in India, Bangladesh, and Pakistan. Any attempt to pursue this policy would impose intolerable burdens on the police and the courts.

Compulsory sterilization, either of all who have had their quota of children or of all who exceed their quota, is much favored by those who regard themselves as hardheaded realists. It is morally unavailable as violating the rights to bodily integrity, to respect as persons, and more generally, as inhumane. When applied to all who have had their quota of children it is morally objectionable as seeking to prevent the commission of a crime even before there are any grounds for believing that a crime is being contemplated. If sterilization is restricted to those who are so punished for having an excessive number of children it is to be objected to as a form of mutilation punishment. In enlightened, liberal societies, there is a general unwillingness to seek to prevent criminals, even habitual criminals such as voyeurs, pickpockets, rapists, from repeating their crimes by mutilating them, blinding the voyeur, chopping off the hand of the pickpocket, castrating the rapist. This is because this kind of punishment is seen to be unjust, often excessive, and where not excessive, morally inappropriate. In the case of sterilization it could seriously harm, pyschologically if not physically, the man or woman who is so punished. It represents an attempt to prevent the crime in advance, even though we cannot know who are the potential recidivists, while justice and respect for persons dictate that we restrict punishment to actual offenders.

It may be argued that some of the objections to general, compulsory sterilization and to sterilization as punishment would lose some of their force if sterilization were to be reversible, that many who now resist sterilization because they may have children who die, or because they may remarry and wish to have children with their new spouse, would not resist reversible sterilization. It is true that some persons may morally object less, or less strongly, to compulsory sterilization if it were reversible, but the main objections would stand.

Further, as noted in Part II, much of the appeal of sterilization as a method of birth control lies in its irreversible character. If those who were sterilized could have the operation reversed at will, it would cease to be the powerful weapon it is believed to be for reducing birth rates.

Compulsory sterilization, whether of all who have had their quota of children or of those who exceed their quota by one, is obviously completely unacceptable on practical, political grounds. Unless the persons to be treated were unusually docile, given that for women the operation involved is one that is both unpleasant and carries the risks that significant operations carry, and that for men it is one that many would resist for a large variety of reasons not least being the fear of loss of "manhood," the state would be faced with using its police force to bring hundreds of thousands of persons every day to clinics to have operations performed on them as unwilling "patients" by, presumably, a conscripted body of doctors. Ecological moralists commonly favor the use of this method of birth control in high-birth-rate, high-population countries, India, Bangladesh, Pakistan, Mexico, Brazil. In those countries it would involve forcibly taking tens of thousands of persons daily to clinics, where many had religious, moral, psychological and personal objections to the operation. The effects on the police and doctors and on respect for the police could not be other than bad. Corruption obviously would occur. And, with the lack of public cooperation, such a scheme would soon collapse. and it would indirectly bring about many other deaths due to the use of necessary hospital facilities.

Abortion is one of the most important, most widely used methods of birth control. Yet, as argued in Part II, there are serious grounds for questioning the morality of using abortion as a method of birth control. There are morally compelling reasons why compulsory abortion must be rejected as a method of population control. Few issues in morality are so obscure, so undecidable in a decisive way, as is that of the morality of abortion. Many moralists, and not simply Roman Catholics, see abortion to be morally evil, as tantamount to murder. Hence to force persons to undergo abortions against their wills is to force them to be unwilling parties to what they see to be akin to murder. They would see themselves as morally obliged to resist the abortion. Otherwise they would be acting as accessories to quasi murder, and incur the moral guilt of being such. Thus compulsory abortion would involve forcing resisting, highly motivated, moral persons to come to hospitals and clinics forcibly to submit to the unwanted killing of their fetuses. The violation of rights, the lack of respect for persons as moral agents and for their integrity as persons and the integrity of their bodies, would be vast if such a method of population control were to be successful. With the millions, possibly hundreds of millions, of abortions such a policy would involve, it is inevitable that many women will die as a result of the abortions and their side effects. Only the gravest of moral reasons could justify such a massive violation of basic human rights.

The morally least offensive of the coercive methods of birth control is that of late marriage backed by social sanctions and legal disabilities against those occupying the status of unmarried parent and illegitimate child. Much is made by some writers of the success of late marriage in Ireland and China in reducing the birth rate, as if this approach were one that did not depend on coercion. In fact, it is an approach that rests heavily on coercion, on coercion of kinds that involve grave injustices and many evils, such that most morally thoughtful persons not preoccupied with the issue of the population crisis are greatly in favor of the recent trends of lessening this coercion, the sanctions, disabilities, and evils, even though this has resulted in many more illegitimate births. The sanctions, disabilities, and coercion are vital elements if late marriage is to contribute to lowering the birth rate and thereby solving the population problem. Late marriage contributes to lowering the birth rate only so long as persons accept the view that births should be confined to marriage. Sanctions, social pressure of various kinds, are applied to foster and reinforce this view. The unmarried mother is ostracized; her child is accorded a much inferior legal and social status compared with the legitimate child and is stigmatized a "bastard." With the weakening of religious beliefs and sanctions and changing moral beliefs, late marriage is a much less effective method of checking the birth rate in Western countries than it was a generation ago.

There are, then, serious moral and political objections to all approaches to population control based on the use of the state's coercive powers, other than that of making it a crime to have more than the specified number of offspring. The latter approach would seriously interfere with freedom of choice and action, but it could be justified if there were a real danger of overpopulation and the great evils that overpopulation would bring into being.

The population crisis, if there comes to be one, will not come from population growth in Europe, the United States, or Australasia in the near future, but from population growth in Asia and Latin America, and, to a much lesser extent, Africa. It could well be that for various reasons, national pride and ambition, indifference, or religious conviction, various of the nations in these areas will not act, or will not act effectively, to check population growth. If their inaction leads to world overpopulation, serious moral issues as to the duties and rights of other peoples will arise. G. Hardin, in "The Tragedy of the Commons" (1968) and in "Lifeboat Ethics: The Case Against Helping the Poor" (1977), has argued that the wealthy nations have the right and duty to withhold aid from the peoples of such nations. Many have echoed Hardin's general thesis. It would appear that the most popular and seemingly morally most respectable version is that the wealthy nations have the moral right to withhold food and medical aid from the starving millions of such nations, unless and until they adopt effective programs of population control, where the wealthy nations decide what is a satisfactory program.

Hardin and those who share this general view see it as being applicable and very relevant today. A legitimate, initial, first response to such a contention is that it is inadvertent racism, it is requiring peoples of other nations and other races to submit to coercion of a kind we should not entertain imposing on our own nationals or on members of our own race. We know that no democratic government would entertain imposing compulsory abortion, sterilization, or contraception on the peoples of the U.S., the U.K., France, Germany, Australia. Yet we are claimed to have the right, by withholding food, and seemingly, by implication, also medicines and other essentials, from Asians, Latin Americans, and Africans to let innocent persons die. A second response is that such conduct grossly violates the rights of those left to starve to death. The rights to life and health are rights of recipience, rights to aids and facilities, to what is necessary to live and be healthy. This view underpins the ethic of the welfare state, but it is a truth seen also by many liberals prior to this century, including no less a philosopher than John Locke. A third response rests in noting the strange, indefensible view of property rights that underlies this view. Even if obtained justly, property carries with it responsibilities and duties of stewardship. Roman Catholic moralists have long acknowledged this, as when they noted that the starving man has the moral right to take from the wealthy that which is necessary for survival, and that in so acting, he is not guilty of theft. Further, the property of the wealthy nations is property that has been acquired in many ways, many of them unjust ways. Much of it rests on past exploitation, wars of conquest, slavery, and the like. Today, much results from extremely unjust trading associations such as the European Economic Community, the raising of restrictions on trade by way of tariffs and quotas to prevent free, fair, competitive trade between all countries of the world that would benefit the poorer nations and render many of them not in need of "charity." Today, we are confronted with the scandal of countries with starving millions they cannot feed having to export food that they produce and desperately need to wealthy countries which already have a surplus of food, and this in order to obtain other essentials. Until the world rectifies past injustices and adopts fair trading practices, property rights of nations cannot be viewed as having a solid moral basis. Even when they come to have a solid moral basis, property rights will still carry with them duties toward the needy.

Hardin's proposals in respect of today's world are completely morally unacceptable. However, they do point to a real problem should the world come to be in danger of being overpopulated. Suppose the world were to come to adopt fair and free trading practices, suppose past injustices were rectified, and suppose that even when it came to be apparent that the world was becoming overpopulated, some high-population, impoverished countries continued to overproduce such that they could not provide for their populations. Would the wealthier nations have only duties to respect the rights to life and health of the destitute in such countries, or would they have rights to act to enforce observance of their moral obligations not to produce children for whom care could not adequately be provided on those who ignored their moral duties?

I suggest that no nation can have the right to deny the help the innocent, needy children require, nor to intervene to force the relevant state to act in the matter. Only a legitimately constituted world political authority can have that right. No nation by virtue of superior wealth or power has the moral right to force another nation to carry out its duties toward its population and to deny to innocent human beings what is necessary for the enjoyment of their basic rights. A world political authority can have the former right. Thus it is that a consideration of one of the problems that makes up "the ecological crisis" leads to the conclusion that satisfactory resolution of that crisis can only be realized in and through a world political authority. No serious proposals relating to how such an authority is to be brought into being have been explored, let alone seriously developed by those who most stress the imminence and gravity of the ecological crisis.

15

The Politics of Ecological Reforms:
The Liberal Democratic Social Order

The ecological reforms claimed to be necessary because of the ecological crisis, reforms to secure preservation, conservation of resources, reduction of pollution, stabilization of population, are far-reaching both in their interference with personal rights and in the very considerable powers that would need to be exercised by nation-states and a world political authority. Many of the reforms, even if seen to be necessary, would not be popular. Among ecological moralists and political theorists, there is a considerable distrust of democracy and democratic processes, and this is tied to the likely unpopularity of the necessary ecological reforms. It is also related to the belief that swift, decisive, effective action is needed in introducing the reforms, the crisis being seen to be imminent. Democracies are seen as not lending themselves to swift, decisive, unpopular action. People, including legislators, have to be persuaded, bureaucrats have to be made to be cooperative, laws have to be framed, passed, implemented, and all this is a context in which effective opposition is given institutional recognition and backing. Other ecological reformers despair of liberal democracies because they see them as resting on self interest and on the reconciliation or compromise of clashes of interests, when ecological reforms are seen to call for farsighted vision and for concerns that go far beyond selfish interests. In democracies, legislators have to be elected and are subject to reelection. How can persons hope to be elected and reelected if they support unpopular measures, measures not related to furthering the interests of their electorate? All the measures seen to be necessary by those who speak of an "ecological crisis," measures to secure preservation and check resource depletion, pollution, and population growth, can be very unpopular with a democratic electorate. They may necessitate less convenient, changed lifestyles, many controls and restrictions, increased costs, additional taxes, basic interference with what now are seen as being matters of basic rights. How then can democratic legislators hope to bring about the reforms that are claimed to be necessary, and still successfully retain office? This leads to the thought that perhaps ecological reformers may need to look for political solutions through totalitarian political organizations, in ecological totalitarianism, ecological ruling elites, ecological philosopher-kings.

Many responses to such suggestions are possible. Democracies and the citizens thereof will not embrace unpopular measures as solutions to problems if they do not believe the problems to be real problems. Unless and until convincing evidence is adduced that there are genuine ecological problems calling for urgent action, democratic citizens will not vote for unpopular measures that deprive them of their liberties and of valued goods. However, when democrats see that there is an evident need for otherwise unpopular measures, they readily and cooperatively accept them, as they did during World War II. Second, while self-interest and the resolution of clashes of interests play a major role in day-to-day democratic politics, it is a distortion of the role democracies play to suggest that they are solely concerned with resolution of clashes of interests. Democracies are concerned about values, rights, ideals. The welfare state was not the outcome of a resolution of clashes of interests of those possessed of political power; it was rather the result of an acceptance of values, rights, ideals. The growing awareness by wealthy nations of the nature and extent of their duties in respect of the poorer nations is again something that is not a product of clashes of selfish interests but a seeing beyond mere self-interest to moral rights and duties.

Democracies are capable of swift, decisive action when there is evident need for such. In democracies, there has already been a good deal of swift, decisive action in ecological matters, much more than in totalitarian states, where the facts have been clear and the need for quick decisive action evident. As open societies, democracies allow the facts to be known and disseminated. Many of the important ecological measures that today are being implemented are being implemented in democracies because they allow free discussion, dissemination of information, agitation for reform, and protests when reforms are not forthcoming.

By contrast, if we consider actual totalitarian states, China, Chile, the USSR, Argentina, the dictatorships of Africa and the Arab world, we find that they are far from ecologically minded and far from being the efficient, centrally planned states that idealistic ecological reformers imagine centrally planned states to be. China and the USSR are among the worst ecological offenders in regard to pollution; they have shown little ecological awareness or sensitivity in interferences with nature. They are less wasteful of resources than are the affluent democracies but the reasons for this are not those of ecological concern. Chile and Argentina are states that illustrate how dictatorships that come to power under the guise of safeguarding human values may themselves come to constitute greater threats to them than did the governments from which they claimed to be saving their communities. Ecologically sponsored dictatorships might reasonably be expected to lose sight of their source and roots. Such states bring out the very considerable dangers that arise once states exercise the right to engage in mass manipulation and employ arbitrary power to realize their chosen ends. No existing dictatorship or totalitarian state provides any grounds for believing that the dangers inherent in totalitarian government are not at least as great as are those that result from ecological problems. Indeed, there are grounds for believing that totalitarian states, through their various policies in respect of war, nuclear weapons, and germ warfare, may precipitate a real ecological crisis.

Even if it were to be shown that ecological measures could safely be entrusted to a totalitarian state, the problem would remain of securing the successful setting up of the appropriate ecological dictatorships. What ecological political theorists who are attracted to a solution through an ecological dictatorship need to discuss is something akin to a Platonic republic but ruled by ecologically wise, morally infallible rulers who know how to bring humanity to live in harmony with its environment. The problems in the way of showing that such an ecologically elitist government is possible and feasible include all those that confronted Plato, and many more besides. Plato was concerned to secure simply wise, informed, incorruptible, disinterested rulers. The ecological elitist must write in the additional requirement that, besides possessing political wisdom, the rulers must also possess ecological wisdom. The ecological Platonist will seek to secure and maintain this wisdom in a closed society, one in which free discussion is certain to be curtailed. Yet it is in the open societies of the world that there has been the growing, deepening awareness and concern for ecological issues, and in which the problems that result from oversimplified, harmful, dangerous responses to the problems are exposed and revealed as such, and to very good effect. It is improbable that such full, frank, free, critical discussion of the issues would be permitted in an ecological totalitarian state. Quite apart from the dictator/s, the bureaucrats of an ecological republic would be unlikely to have much patience with those who stressed the complexities of the issues, the difficulties to be met, the inadequacies of simplistic, unintegrated solutions. Further, just as Plato's republic is at best an unrealizable utopian ideal, one we should do better to put aside when realistically planning the good society, so the ecological republic of the ecological utopian dreamer is best put aside as dangerous and irrelevant, and attention concentrated on realistic, practical political action.

An ecological totalitarianism would need to educate and morally improve its citizens in ways exactly parallel to those in which ecologically minded democratic states would have to educate their citizens; otherwise they would become the worst kinds of police states. They would be police states that imposed on the police, the criminal law, and the courts burdens that they could not possibly successfully carry. Even so, there would have to be great invasions of privacy, extensive use of fear as an instrument of social control. The little likelihood of greater success than that to be realized in a liberal democracy, provides no ground for jeopardizing our enjoyment of our human rights, in particular our right to liberty.

There is also the problem that Plato never adequately solved in respect of his republic, namely, that of how his philosopher-dictators would be brought to power. It would seem not to be possible to achieve this by democratic means, and this for two reasons. First, is that what makes for difficulty in getting ecologically minded dictators elected is that which also makes for the problem of securing the election of ecologically minded democratic parties, prime ministers, presidents. Second, democratic elections are very chancy affairs, and very often they result in leaders who pursue policies of which they have given no notice, or who change their policies, often rightly, due to force of circumstance, where the circumstance may be economic, political, or military. The alternative approach, that of seeking to bring to power by illegal, undemocratic military means the ecological dictators who are desired, imposing them by force on the peoples of the world, is obviously unreliable and unrealistic. We may simply destroy what we have and still not succeed in setting up genuinely ecologically orientated dictatorships. Such persons might well prove to be completely incompetent as ecological dictators; they may prove to be dictators who bring the world community to ruin by imposing oversimplified, ill-thought-out solutions to pseudo-problems that they see to be real problems. The political naiveté of those concerned with ecological issues, especially those of a totalitarian turn of mind, gives us every reason to distrust their political competence.

In brief: The only realistic, feasible avenue to ecological political reform is through the political institutions of an open society that respects human rights. Those of us who are fortunate enough to live in liberal democratic states can hope to have a much greater measure of success than can those who live in closed societies, even those that sail under the flag of ecological concern, working toward solutions of ecologically based political problems.

Bibliography

Aiken, W., and La Follette, H., eds. 1977. *World Hunger and Moral Obligation*. Englewood Cliffs, N.J.: Prentice-Hall.

Anderson W., ed. 1970. *Politics and Environment*. Pacific Palisades. Calif.: Goodyear.

Aquinas, Saint Thomas. *Summa Theologiae (Summa Theologica)* 1947. Trans. Dominican Fathers. New York: Beniger Brothers.

Aristotle. *Nicomachean Ethics*. Trans. W. D. Ross, 1915. Oxford: Oxford University Press.

Blackstone, W. T., ed. 1974a. *Philosophy and Environmental Crisis*. Athens, Ga.: University of Georgia Press.

———. 1974b. "Ethics and Ecology." In *Philosophy and Environmental Crisis*, ed. W. T. Blackstone, pp. 16–42. Athens: University of Georgia Pres.

———. 1979. "The Search for an Environmental Ethic." In *Matters of Life and Death*, ed. T. Regan, pp. 299–331. New York: Random House.

Boughey, A. S. 1975. *Man and the Environment*. New York: Macmillan.

Boulding, K. E. 1966. "The Economics of the Coming Spaceship Earth." In *Environmental Quality in a Growing Economy*, ed. H. Jarrett, pp. 3–14. Baltimore: John Hopkins University Press. Reprinted in *Pollution, Resources and the Environment*, ed. A. C. Enthoven and A. M. Freeman, pp. 14–24. New York: Norton, 1973.

Carson, R. 1962. *Silent Spring*. Boston: Houghton Mifflin.

Cole, H. S. D.; Freeman, Johada M.; and Pavitt, K. L. R., eds. 1973. *Thinking about The Future*. London: Chatto and Windus.

Commoner, B. 1967. *Science and Survival*. New York: Viking.

———. 1972. *The Closing Circle*. New York: Bantam.

Davis, K. 1967. "Population Policy: Will Current Programs Succeed?" In *Man and the Environment*, ed. W. Jackson. Dubuque, Iowa: W. C. Brown, pp. 180–99. Originally published in *Science* 158: 730–39.

———. 1973. "Zero Population Growth: The Goal and the Mean." In *The No-Growth Society*, ed. M. Olson and H. L. Landsberg, pp. 15–30. New York: Norton.

Disch. R., ed. 1970. *The Ecological Conscience*. Englewood Cliffs, N.J.: Prentice-Hall.

Dubos, R. 1965. *Man Adapting*. New Haven: Yale University Press.

———. 1970. *Man, Medicine and Environment*. Harmondsworth: Penguin.

———. 1973. *A God Within*. London: Angus and Robertson.

Du Boff, R. B. 1974. "Economic Ideology and the Environment." In *Man and the Environment Ltd.*, ed. H. G. T. Van Raay and A. E. Lugo, pp. 201–20. The Hague: Rotterdam University Press.

Ehrlich, P. R. 1968a. "World Population: A Battle Lost?" In *Politics and Environment*, ed. W. Anderson, pp. 13–21. Pacific Palisades, Calif.: Goodyear. Originally published in *Stanford Today*, 1968 Series 1, 22.

———. 1968b. *The Population Bomb*. New York: Ballantine.
Ehrlich, P. R., and Ehrlich, A. H. 1970. *Population, Resources, Environment*. San Francisco: W. H. Freeman.
Ehrlich, P. R.; Ehrlich, A. H.; and Holdren, J. 1973. *Human Ecology: Problems and Solutions*. San Francisco: W. H. Freeman.
Ellul, J. 1964. *The Technological Society*, trans. J. Wilkinson. New York: Knopf.
Enthoven, A. C., and Freeman, A. M., eds. 1973. *Pollution, Resources and the Environment*. New York: Norton.
Falk, R. A. 1971. *This Endangered Planet*. New York: Vintage.
Feinberg, J., ed. 1973. *The Problem of Abortion*. Belmont, Mass.: Wadsworth.
———. 1974. "The Rights of Animals and Unborn Generations." In *Philosophy and Environmental Crisis*, ed. W. T. Blackstone, pp. 43–68. Athens: University of Georgia Press.
Forrester, J. W. 1971. *World Dynamics*. Cambridge, Mass.: Wright-Allen.
Frey, R. G. 1977. "Animal Rights." *Analysis* 37: 186–89.
———. 1980. *Interests and Rights*. Oxford: Oxford University Press.
Gerber, J. F. 1974. "Agricultural Technology and Food for a Hungry World." In *Man and the Environment Ltd.*, ed. H. G. T. Van Raay and A. E. Lugo, pp. 77–102. The Hague: Rotterdam University Press.
Goodpaster, K. E., and Ayre, K. M., eds. 1979. *Ethics and Problems of the 21st Century*. Notre Dame, Ind.: Notre Dame University Press.
Hardin, G. 1968. "The Tragedy of the Commons." *Science* 162: 1243–48.
———. 1969. "Not Peace, but Ecology." *Diversity and Stability in Ecosystems*, Symposium, May 26-28, 151–58. Biology Department, Brookhaven National Laboratory, Upton, N.Y., BWL 50175 (650).
———. 1972. *Exploring New Ethics for Survival*. New York: Viking.
———. 1975. Foreword in *Living in the Environment*, ed. G. Tyler Miller. Belmont, Mass.: Wadsworth.
———. 1977. "Lifeboat Ethics: The Case Against Helping the Poor." In *World Hunger and Moral Obligation*, ed. W. Aiken and H. La Follette, pp. 14–21. Englewood Cliffs, N.J.: Prentice-Hall.
Harman, G. 1977. *The Nature of Morality*. New York: Oxford University Press.
Hart, H. L. A. 1955. "Are There Any Natural Rights?" *Philosophical Review* 64: 175–91.
Hobbes, T. 1651. *The Leviathan*. 1914. London: Dent.
Jackson W., ed. 1971. *Man and the Environment*. Dubuque, Iowa: W. C. Brown.
Jevons, W. S. 1865. *The Coal Question: An Inquiry Concerning the Progress of the Nation and the Probable Exhaustion of Our Coal Mines*. 1965. New York: Augustus M. Kelley.
Kraft, M. E. 1977. "Political Change and the Sustainable Society." In *The Sustainable Society*, ed. D. C. Pirages, pp. 173–96. New York: Praeger.
Lamb, R. 1980. "On Ecological Ethics and Its Justification." In *Environmental Philosophy*, ed. D. S. Mannison, M. A. McRobbie, and R. Routley, pp. 88–95. Canberra: Australian National University.
Leopold, Aldo C. 1933. "A Conservation Ethic." *Journal of Forestry* 31: 634–43.
———. 1934. "Conservation and Economics." *Journal of Forestry* 32: 537–44.
———. 1966. *A Sand County Almanac*. New York: Oxford University Press.
Locke, J. 1690. *Second Treatise of Civil Government*. London: Dent. 1924.
Lyons, D. 1979. "Are Luddites Confused?" *Inquiry* 22: 381–403.
Macaulay, T. B. 1830. "Southey's Colloquies on Society." *Edinburgh Review*. Reprinted in *Thomas Babington Macaulay: Selected Writings*, ed. J. Clive and T. Pinney, pp. 34–78. Chicago: University of Chicago Press, 1972.
McCloskey, H. J. 1965. "Rights." *Philosophical Quarterly* 15: 115–27.
———. 1969. *Meta-Ethics and Normative Ethics*. The Hague: Martinus Nijhoff.
———. 1975. "The Right to Life." *Mind* 84: 403–25.
———. 1979. "Moral Rights and Animals." *Inquiry* 22: 23–54.
———. 1980a. "Ecological Ethics and Its Justification: A Critical Appraisal." In *Environmental Philosophy*, ed. D. S. Mannison, M. A. McRobbie, and R. Routley, pp. 65–87. Canberra: Australian National University.

————. 1980b. "Ecological Values the State and the Right to Liberty." *Pacific Philosophical Quarterly* 61: 212–31.

McHarg, I. L. 1971. *Design with Nature.* New York: Doubleday.

Maddox, J. 1972. *The Doomsday Syndrome.* London: Macmillan & Co.

Malthus, T. R. 1798. *An Essay on the Principle of Population as It Affects the Future Improvement of Mankind.* Ed. P. Appleman. New York: Norton, 1976.

Mannison, D. S.; McRobbie, M. A.; and Routley, R., eds. 1980. *Environmental Philosophy.* Canberra: Philosophy Department, Research School of Social Science, Australian National University.

Marx, K. 1860–1863. "Theories of Surplus Values." In *Marx and Engels on Mathus,* ed. by R. L. Meek and trans. by R. L. Meek and D. L. Meek. New York: New York International. 1954.

Meadows, D. H.; Meadows, D. L.; Randers, J.; and Brehrens, W. W. 1974. *The Limits to Growth.* London: Pan.

Mesthene, E. G. 1968. "How Technology Will Shape the Future." *Science* 161: 135–143.

Mill, J. S. 1848–1871. *Principles of Political Economy. Collected Works of John Stuart Mill.* Vols. 2 and 3. Toronto: Toronto University Press, 1965.

————. 1859. *On Liberty. Collected Works of John Stuart Mill,* Vol. 18. Toronto: Toronto University Press, 1977.

Miller, G. T., ed. 1975. *Living in the Environment.* Belmont, Mass.: Wadsworth.

Mishan, E. J. 1967. *The Costs of Economic Growth.* London: Staples.

————. 1970. *Technology and Growth.* New York: Praeger.

Morrison, J. F. 1974. "Man, Organization, and the Environment." In *Man and the Environment,* ed. H. G. T. Van Raay and A. E. Lugo, pp. 177–200. The Hague: Rotterdam University Press.

Naess, A. 1973. "The Shallow and the Deep, Long-Range Ecology Movement. A Summary." *Inquiry* 16: 95–100.

Nelson, L. 1956. *A System of Ethics.* Trans. N. Gutterman. New Haven: Yale University Press.

Oakeshott, M. 1962. *Rationalism in Politics.* London: Methuen.

Odum, E. P. 1974. "Environmental Ethic and the Attitude Revolution." In *Philosophy and Environmental Crisis,* ed. W. T. Blackstone, pp. 10–15. Athens: University of Georgia Press.

Odum, H. T. 1971. *Environment, Power, and Society.* New York: Wiley Interscience.

Olson, M., and Landsberg, H. L., eds. 1973. *The No-Growth Society.* New York: Norton.

Overbeek, J. 1974. *History of Population Theories.* Rotterdam: Rotterdam University Press.

Paddock, W., and Paddock, R. 1967. *Famine-1975: America's Decision: Who Will Survive.* Boston: Little, Brown. (*Famine—1975!* London: Weidenfeld & Nicolson.)

————. 1970. "Proposal For the Use of American Food: 'Triage.' " In *Politics and Environment,* ed. W. Anderson, pp. 34–36. Pacific Palisades, Calif.: Goodyear.

Parry, H. B., ed. 1974. *Population and Its Problems.* Oxford: Clarendon Press.

Passell, P.; Roberts, M. J.; and Ross, L. 1972. Review of *The Limits To Growth. New York Times Book Review,* 2 April 1972. Reprinted in *Pollution, Resources and the Environment,* ed. A. C. Enthoven and A. M. Freeman, pp. 230–34. New York: Norton.

Passmore, J. 1974. *Man's Responsibility For Nature.* London: Duckworth.

Pirages, D. C., ed. 1977. *The Sustainable Society.* New York: Praeger.

Pirages, D. C., and Ehrlich, P. R. 1974. *Ark II.* San Francisco: W. H. Freeman.

Plato. *The Republic.* Trans. D. Lee, 1955. Harmondsworth: Penguin.

Pope Leo XIII. 1880. *Arcanum Divinae.*

Pope Pius XI. 1930. *Casti Connubii.*

Pope Paul VI. 1968. *Humanae Vitae.*

Regan, T. 1975. "The Moral Basis of Vegetarianism." *Canadian Journal of Philosophy* 5: 181-214.

————. 1976a. "McCloskey on Why Animals Cannot Have Rights." *Philosophical Quarterly* 26: 251-57.

————. 1976b. "Feinberg on What Sorts of Beings Can Have Rights." *Southern Journal of Philosophy* 14: 485-98.

———. 1977. "Frey on Interests and Animal Rights." *Philosophical Quarterly* 27: 335–37.

———. 1979a. "An Examination and Defence of One Argument Concerning Animal Rights." *Inquiry* 22: 189–219.

———, ed. 1979b. *Matters of Life and Death.* New York: Random House.

Regan, T., and Singer, P., eds. 1976. *Animal Rights and Human Obligations.* Englewood Cliffs, N.J.: Prentice-Hall.

Rodman, J. 1977. "The Liberation of Nature?" *Inquiry* 20: 83–131.

Rolston, Holmes III. 1974–1975. "Is There an Ecological Ethic?" *Ethics* 85: 93–109.

Ross, W. D. 1930. *The Right and The Good.* Oxford: Clarendon Press.

———. 1939. *Foundations of Ethics.* Oxford: Clarendon Press.

Routley, R. 1973. "Is There a Need for a New, an Environmental Ethic?" *Proceedings of the XVth World Congress of Philosophy,* Vol. 1, 205-10.

Routley, R., and Routley, V. 1980. "Human Chauvinism and Environmental Philosophy." In *Environmental Philosophy,* ed. D. S. Mannison, M. A. McRobbie, and R. Routley, pp. 96-189. Canberra: Australian National University.

Routley, V. 1975. Critical Notice, J. Passmore's *Man's Responsibility For Nature. Australasian Journal of Philosophy* 53: 171-85.

Sauvy, A. 1975. *Zero Growth.* Oxford: Basil Blackwell.

Sax, K. 1955. *Standing Room Only.* Boston: Beacon.

Singer, P. 1975. *Animal Liberation.* New York: New York Review of Books Press.

Stone, C. D. 1972. *Should Trees Have Standing: Toward Legal Rights For Natural Objects.* Los Altos, Calif.: William Kaufman.

Teichman, J. 1978. *The Meaning of Illegitimacy.* Cambridge, England: Englehardt Books. Rev. ed.: 1982. *Illegitimacy.* Ithaca: Cornell University Press.

United Nations Fund for Population Activities. 1981. *The State of World Population 1981.*

Van Raay, H. G. T., and Lugo, A. E., eds. 1974. *Man and the Environment Ltd.* The Hague: Rotterdam University Press.

Ward, B. 1979. *Progress for a Small Planet.* New York: Norton.

Watson, R. A. 1971. "Human Dignity and Technology." *The Philosophy Forum* 9: 221–45.

Watson, R. A., and Smith, P. M., 1970. "The Limit: 500 Million." *Focus/Midwest* 8: 25–27.

Williams, Roger. 1644. *The Bloudy Tenent of Persecution, The Complete Writings of Roger Williams,* Vol. 3. New York: Russell & Russell. 1963.

Index